Statistics for tech

Christopher Chatfield

Penguin Books

Penguin Books Ltd, Harmondsworth,
Middlesex, England
Penguin Books Inc., 7110 Ambassador Road,
Baltimore, Md 21207, U.S.A.
Penguin Books Australia Ltd, Ringwood,
Victoria, Australia

First published 1970
Copyright © Christopher Chatfield, 1970

Filmset by J. W. Arrowsmith Ltd, Bristol
in Monophoto Times New Roman,
and made and printed in Malta by
St Paul's Press Ltd

Penguin Education
Studies in applied statistics
General Editor: I. R. Vesselo

Statistics for technology

Christopher Chatfield

Contents

6 Contents

7   Contents

'It's perfectly intelligible,' the captain said, in an offended
tone, 'to anyone that understands such things.'

Preface

The purpose of this book is to acquaint the reader with the increasing number of applications of statistics in engineering and the applied sciences. It can be used either as a text for a basic course in statistics or for self-tuition. The mathematics is kept as simple as possible. An elementary knowledge of calculus plus the ability to manipulate algebraic formulae are all that is required.

While writing this book I have been conscious of the fact that there are already a number of good comprehensive books for engineers on basic statistical theory. My goal has been to introduce this basic theory without getting too involved in mathematical detail, and thus to enable a larger proportion of the book to be devoted to practical applications. Because of this some results are stated without proof, where this is unlikely to affect the student's comprehension. However, I have tried to avoid the cook-book approach to statistics by carefully explaining the basic concepts of the subject, such as probability and sampling distributions; these the student must understand. The worst abuses of statistics occur when scientists try to analyse their data by substituting measurements into statistical formulae which they do not understand.

This book would not be complete without a few remarks on the ever-increasing use of digital computers. These machines have had a tremendous effect on scientific research. Firstly, they enable the experimenter to investigate more complicated models than would be practicable if the computation had to be done on a desk calculating machine. Secondly, they have encouraged research in many new directions where the computational problems had previously been formidable. Two statistical examples of this are multivariate analysis and time-series analysis. Computers are also extensively used for much routine data analysis. A package of statistical programs is now available with most computers for performing such tasks as multiple regression and the analysis of variance. Nevertheless the reader is advised to use a desk calculating machine to work through the problems given in this book, as this should give him a deeper understanding of the subject and enable him to use the computer for more complicated problems.

Originally this book was to be a joint venture between myself and Dr R. B. Abernethy, who is an engineer with Pratt and Whitney Aircraft. Unfortunately Dr Abernethy was forced to withdraw from the project because of pressure of work. Nevertheless I am very grateful to him for his valuable assistance while the book was taking shape, and also for early drafts of Chapters 3 and 9. Of course any errors or omissions which remain are entirely my responsibility.

I would also like to thank my colleagues at McGill University for reading parts of the book and offering helpful criticisms and also to record my appreciation to Mrs R. Rigelhof and Miss H. Schroeder for undertaking the task of preparing the manuscript.

I am indebted to Biometrika trustees for permission to publish extracts from *Biometrika Tables for Statisticians* (Appendix B, Table 6) and to Messrs Oliver and Boyd Ltd, Edinburgh for permission to publish extracts from *Statistical Tables for Biological, Agricultural and Medical Research* (Appendix B, Table 7). I am also indebted to Penguin Education for providing Tables 1, 2, 3, 4 and 5 in Appendix B.

The quotations are of course from the works of Charles Lutwidge Dodgson, better known as Lewis Carroll.

The book is divided into three sections; Part one is the introduction, Part two deals with statistical theory and Part three considers some applications. The allocation of the material to the different sections is somewhat arbitrary; for example regression problems, which are discussed in Part two, would probably be equally at home in Part three. Among the applications which are considered in Part three are the planning, design and analysis of experiments, quality control and life-testing. The reader who is specifically interested in one of these last two topics can proceed directly from Chapter 7 to Chapter 12 or 13, as appropriate.

CHRISTOPHER CHATFIELD

Part one
Introduction

'Surely,' said the governor, 'Her Radiancy would admit
that ten is nearer to ten than nine is – and also nearer
than eleven is.'

Chapter 1
Outline of statistics

Statistics is an increasingly important subject which is useful in many types of scientific investigation. It has become the science of collecting, analysing and interpreting data in the best possible way. Statistics is particularly useful in situations where there is *experimental uncertainty* and is often defined as 'the science of making decisions in the face of uncertainty'. We begin with some scientific examples in which experimental uncertainty is present.

Example 1

The thrust of a rocket engine was measured at ten-minute intervals while being run at the same operating conditions. The following thirty observations were recorded.

999·1	1003·2	1002·1	999·2	989·7	1006·7
1012·3	996·4	1000·2	995·3	1008·7	993·4
998·1	997·9	1003·1	1002·6	1001·8	996·5
992·8	1006·5	1004·5	1000·3	1014·5	998·6
989·4	1002·9	999·3	994·7	1007·6	1000·9

The observations vary between 989·4 and 1014·5 with an average value of about 1000. There is no apparent reason for this variation which is of course small compared with the absolute magnitude of each observation; nor do the variations appear to be systematic in any way. Any variation in which there is no pattern or regularity is called *random variation*. In this case if the running conditions are kept uniform we can predict that the next observation will also be a thousand together with a small random quantity which may be positive or negative.

Example 2

The numbers of cosmic particles striking an apparatus in forty consecutive periods of one minute were recorded as follows.

```
3 0 0 1 0 2 1 0 1 1
0 3 4 1 2 0 2 0 3 1
1 0 1 2 0 2 1 0 1 2
3 1 0 0 2 1 0 3 1 2
```

The observations vary between zero and four, with zero and one
observed more frequently than two, three and four. Again there is
experimental uncertainty since we cannot exactly predict what the
next observation would be. However, we expect that it will also be
between zero and four and that it is more likely to be a zero or a one
than anything else. In Chapter 4 we will see that there is indeed a
pattern in this data even though individual observations cannot be
predicted.

Example 3

Twenty refrigerator motors were run to destruction under advanced
stress conditions and the times to failure (in hours) were recorded as
follows.

104·3	158·7	193·7	201·3	206·2
227·8	249·1	307·8	311·5	329·6
358·5	364·3	370·4	380·5	394·6
426·2	434·1	552·6	594·0	691·5

We cannot predict exactly how long an individual motor will last,
but, if possible, we would like to predict the pattern of behaviour of a
batch of motors. For example we might want to know the over-all
proportion of motors which last longer than one week (168 hours).
This problem will be discussed in Chapter 13.

When the scientist or engineer finishes his education and enters
industry for the first time, he must be prepared to be faced frequently
with situations which involve experimental uncertainty. The purpose of
this book is to provide the technologist with methods for treating these
uncertainties. These methods have proved to be very useful in both
industry and research.

The industrial experiment has some or all of the following characteristics.

(1) The physical laws governing the experiment are not entirely understood.

(2) The experiment may not have been done before, at least successfully, in which case the instrumentation and technique are not fully developed.

(3) There are strong incentives to run the smallest number of the cheapest tests as quickly as possible.

(4) The experimenter may not be objective, as for example when an inventor tests his own invention or when a company tests competitive products.

(5) Experimental results are unexpected or disappointing. (Engineers explain disappointing results as illustrating Murphy's law. The corresponding law for statistics might be phrased 'if two events are equally likely to occur, the worse will happen'.)

(6) Although experimental uncertainty may be present, many industrial situations require decisions to be made without additional testing or theoretical study.

To illustrate how statistics can help at different stages of an experiment, let us assume that we have been given the task of improving the performance of a space pump. The basic design of this machine is illustrated in Figure 1: gas is passing through the pump which is driven by an electric motor.

The first step is to study the current technology. Statistical methods are never a substitute for understanding the physical laws governing

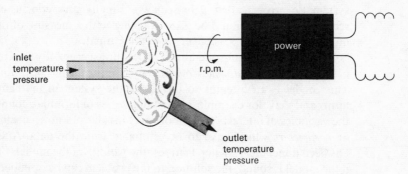

Figure 1   Diagram of a space pump

the problem under consideration; rather statistics is a tool for the technologist in the same sense as differential equations and digital computers.

The second step is to define the *objective* of the test program as precisely as possible. For example we may be interested in improving flow, pressure rise or efficiency, or reducing weight or noise. Alternatively we may be interested in improving the *reliability* of the machine by increasing the time between breakdowns.

The objective being defined, a list of *variables* or *factors* can be made which will vary or be varied during the test program. In the space-pump experiment, the variables include power, inlet temperature, inlet pressure and speed. The test program must be *designed* to find the best way of choosing successive values for the different factors. The problems involved in designing an experiment are discussed in Chapters 10 and 11.

In order to see if the objective is being achieved, we look at the *outputs* or *responses*. In the space-pump example, these include pressure rise, flow, efficiency and reliability. Note that these responses are probably interconnected. For example it may be pointless to increase the flow if, as a result, the efficiency drops or the reliability falls off (that is, the pump breaks down more frequently).

During the experiment, measurements of the factors and responses will be made. The article being tested should be instrumented to obtain the most *precise* and *accurate* data. The problems of instrumentation are discussed in Chapter 9. The analysis of the resulting data should attempt to determine not only the individual effect of each variable on the responses, but also the joint effects of several variables, the *interactions*. The process of *estimating* these effects is called *inference*. These estimates should have known statistical properties such as a lack of systematic error (called *unbiasedness*). Finally any conclusions or recommendations should be given together with a measure of the risk involved and/or the possible error in the estimates.

In order to carry out the above procedure successfully it is usually helpful to set up a *mathematical model* to describe the physical situation. This model is an attempt to formulate the system in mathematical terms and may, for example, consist of a series of formulae connecting the variables of interest. This model may involve unknown coefficients or *parameters* which have to be estimated from the data. After this has been done, the next step is to test the validity of the model. Finally, if the model is sound, the solution to the problem can be obtained from it.

Additional reading

Further discussion of basic statistical principles can be found in many books, including:

HOOKE, R. (1963), *Scientific Inference*, Holden-Day.
MORONEY, M. J. (1956), *Facts from Figures*, 3rd edn, Penguin Books.
REICHMANN, W. J. (1961), *Use and Abuse of Statistics*, Methuen; paperback edn (1964), Penguin Books.

Chapter 2
Simple ways of analysing data

2.1 Introduction

One of the commonest problems facing the engineer is that of summarizing a number of experimental observations by picking out the important features of the data. Only then will it be possible to try and interpret the results. The simple methods of analysing data which are considered in this chapter are widely used and are often given the collective title 'Descriptive Statistics'. It is convenient to describe them here in the introduction rather than at the start of the theoretical section of the book.

It is important to realize from the outset that the observations are usually a *sample* from the set of all possible outcomes of the experiment (sometimes called the *population*). A sample is taken because it is too expensive and time-consuming to take all possible measurements. Statistics are based on the idea that the sample will be 'typical' in some way and that it will enable us to make predictions about the whole population.

The data usually consist of a series of measurements on some feature of an experimental situation or on some property of an object. The phenomenon being investigated is usually called the *variate*. As a preliminary it is worth emphasizing the distinction between *discrete* and *continuous* data. Discrete data can only take a sequence of distinct values and usually consist of counted observations on a series of integers. Thus in Example 2, Chapter 1, the number of cosmic particles emitted in one minute must be $0, 1, 2, 3, \ldots$. The number $1\frac{1}{2}$ cannot be observed. Other examples of discrete variates are the number of successful tests of a missile and the number of days without rain.

On the other hand continuous data are observations which can take any value on a continuous scale. Thus in Example 1, Chapter 1, the thrust of a rocket engine can be any positive number. Other continuous variates are length, weight, force, and time. Engineers and scientists prefer continuous data where possible as it is more informative.

The summarization and presentation of data, whether discrete or continuous, can be accomplished by a combination of pictorial and arithmetic methods.

2.2 Pictorial methods

It is always a good idea to plot the data in as many different ways as possible, as much information can often be obtained just by looking at the resulting graphs.

2.2.1 *The bar chart*

Given discrete data as in Example 2, Chapter 1, the first step is to find the *frequency* with which each value occurs. Then we find, for example, that 'zero' occurs thirteen times but that 'four' only occurs once.

Table 1

Number of cosmic particles	Frequency
0	13
1	13
2	8
3	5
4	1
Total	40

The values in the right hand column form a *frequency distribution*. This frequency distribution can be plotted as in Figure 2, where the height of each line is proportional to the frequency with which the value occurs.

This diagram is called a *bar chart* (or line graph or frequency diagram) and is a very useful visual aid when analysing *discrete* data.

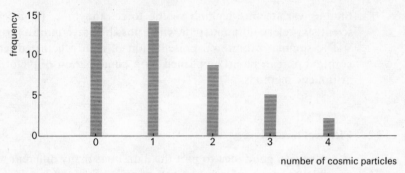

Figure 2   Bar chart of data of Table 1

2.2.2   The histogram

Most people are fairly familiar with the histogram which is a similar type of graph to the bar chart except that it can be used for *continuous* data. It is best illustrated with an example.

Example 1

The heights of 100 students were measured to the nearest inch and tabulated as follows.

Table 2

Height (inches)	Number of students
60–62	6
63–65	15
66–68	40
69–71	30
72–74	9
Total	100

The data were divided into five groups as shown and the frequencies with which the different groups occur form a frequency distribution. The data can be plotted as a histogram in which rectangles are constructed with areas proportional to the frequency of each group.

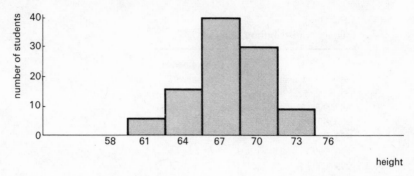

Figure 3   Histogram of data of Table 2

How to draw a histogram

(1) Allocate the observations to between five and twenty *class intervals*. In Example 1 (60–62) inches is a class interval.

(2) The *class mark* is the midpoint of the class interval. All values within the interval are considered concentrated at the class mark.

(3) Determine the number of observations in each interval.

(4) Construct rectangles with centres at the class marks and areas proportional to the class frequencies. If all the rectangles have the same width then the heights are proportional to the class frequencies.

The choice of the class interval and hence the number of intervals depends on several considerations. If too many intervals are used then the histogram will oscillate wildly but if too few intervals are used then important features of the distribution may be overlooked. This means that some sort of compromise must be made. As the number of observations is increased the width of the class intervals can be decreased as there will be more observations in any particular interval.

Example 2

Plot a histogram of the data given in Example 1, Chapter 1. The smallest thrust observed is 989·4 and the largest 1014·5. The difference between them is about twenty-five units so that three units is a reasonable class interval. Then we will have about ten class intervals. Group the observations into the ten intervals as in Table 3.

Table 3
Frequency distribution of thrust

Class interval	Number of observations
987–990	2
990–993	1
993–996	3
996–999	5
999–1002	7
1002–1005	6
1005–1008	3
1008–1011	1
1011–1014	1
1014–1017	1

If an observation falls exactly at the division point (for example 990·0) then it is placed in the lower interval. Note that if we take one unit to be the class interval then there will only be five intervals out of twenty-eight with more than one observation and this will give a very flattened histogram, which is very difficult to interpret.

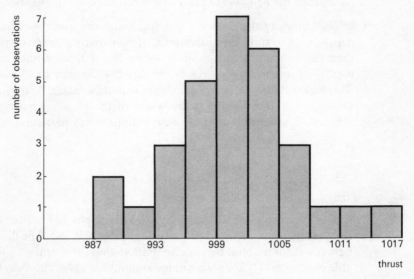

Figure 4   Histogram of thrust data

Histogram shapes. Histograms come in all shapes and sizes. Some of the common shapes are illustrated below.

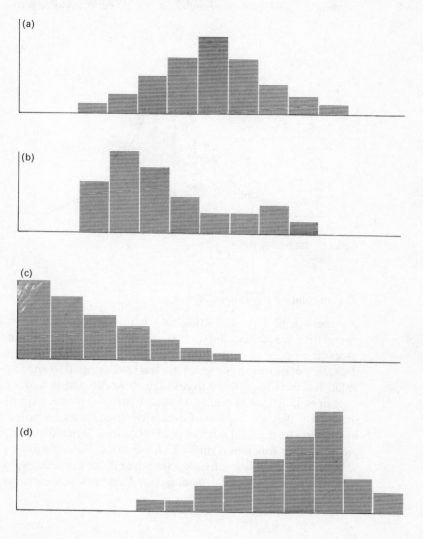

Figure 5   Various histograms
(a) symmetric or bell-shaped
(b) skewed to the right or positively skewed
(c) reverse *J*-shaped
(d) skewed to the left or negatively skewed

Frequency curve. Where there are a large number of observations the histogram may be replaced with a smooth curve drawn through the midpoints of the tops of each box. Such a curve is called a *frequency curve*.

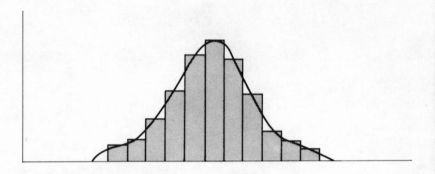

Figure 6   Frequency curve

2.2.3   *The cumulative frequency diagram*

Another useful way of plotting data is to construct what is called a cumulative frequency diagram. If the observations are arranged in ascending order of magnitude, it is possible to find the *cumulative frequency* of observations which are less than or equal to any particular value. It is usually sufficient to calculate these cumulative frequencies at a set of equally spaced points. The cumulative frequencies are easier to interpret if they are expressed as relative frequencies, or proportions, by dividing by the total number of observations. When these values are plotted, a step function results which increases from zero to one.

Interest in cumulative frequencies arises if, for example, we wanted to find the proportion of manufactured items which came up to a certain standard.

Example 3

Plot the cumulative frequency diagram of the data given in Example 1, Chapter 1.

Using Table 3 it is easy to calculate the cumulative frequencies at the divisions between the class intervals.

Table 4

Thrust	Cumulative frequency	Relative frequency
987	0	0·000
990	2	0·067
993	3	0·100
996	6	0·200
999	11	0·367
1002	18	0·600
1005	24	0·800
1008	27	0·900
1011	28	0·933
1014	29	0·967
1017	30	1·000

The relative frequencies have been plotted in Figure 7.

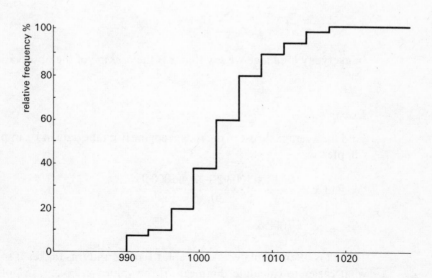

Figure 7   Cumulative frequency diagram of thrust data

2.3 Arithmetical methods

In addition to the graphical techniques, it is often useful to calculate some figures to summarize the data. Any quantity which is calculated from the data is called a *statistic* (to be distinguished from the subject statistics). Thus a statistic is a function of the measurements or observations.

Most simple statistics can be divided into two types; firstly quantities which are 'typical' of the data and secondly quantities which measure the variability of the data. The former are usually called measures of *location* and the latter are usually called measures of *spread*.

2.3.1 *Measures of location*

There are three commonly used measures of location, of which the *mean* is by far the most important.

The mean. Suppose that n measurements have been taken on the variate under investigation, and these are denoted by $x_1, x_2, \ldots, x_n$. The (arithmetic) mean of the observations is given by

$$\bar{x} = \frac{x_1 + x_2 + \ldots + x_n}{n} = \sum_{i=1}^{n} \frac{x_i}{n}. \qquad \textbf{2.1}$$

In everyday language we say that $\bar{x}$ is the *average* of the observations.

Example 4

Find the average thrust of the rocket engine from the data in Example 1, Chapter 1.

$$\text{We find } \bar{x} = \frac{999 \cdot 1 + 1003 \cdot 2 + \ldots + 1000 \cdot 9}{30}$$

$$= 1000 \cdot 6.$$

Data is often tabulated as in Tables 1 and 2 and this makes it somewhat easier to calculate the mean. If the values $x_1, x_2, \ldots, x_N$ of the variate occur with frequency $f_1, f_2, \ldots, f_N$, then the mean is given by the equivalent formula

$$\bar{x} = \frac{f_1 x_1 + f_2 x_2 + \ldots + f_N x_N}{f_1 + f_2 + \ldots + f_N}$$

$$= \frac{\displaystyle\sum_{i=1}^{N} f_i x_i}{\displaystyle\sum_{i=1}^{N} f_i}.$$

Example 5

In Table 1 we have $x_1 = 0$ occurs with frequency $f_1 = 13$. Similarly $x_2 = 1$ occurs with frequency $f_2 = 13$ and so on. Then we find

$$\bar{x} = \frac{(0 \times 13) + (1 \times 13) + (2 \times 8) + (3 \times 5) + (4 \times 1)}{13 + 13 + 8 + 5 + 1}$$

$$= \frac{48}{40}$$

$$= 1{\cdot}2.$$

Thus an average of 1·2 cosmic particles are emitted every minute.

Example 6

When continuous data is tabulated as in Table 2, the sample mean can be estimated by assuming that all the observations in a class interval fall at the class mark.

In this case we find

$$\bar{x} = \frac{(61 \times 6) + (64 \times 15) + (67 \times 40) + (70 \times 30) + (73 \times 9)}{100}$$

$$= 67{\cdot}6$$

$$= \text{mean height of the students.}$$

At this point it is worth repeating the fact that a set of data is a sample drawn from the population of all possible measurements. Thus it is important to realize that the sample mean, $\bar{x}$, may not be equal to the true population mean, which is usually denoted by μ. In Chapter 6 we will see that $\bar{x}$ is an *estimate* of μ. Similarly the other sample statistics described here will also be estimates of the corresponding population characteristics.

The median. This is occasionally used instead of the mean, particularly when the histogram of the observations is *skewed* (see Figure 5). It is obtained by placing the observations in ascending order of magnitude and then picking out the middle observation. Thus half the observations are numerically greater than the median and half are smaller.

Example 7

The wages of twelve workers selected at random from the pay-roll of a factory are

9 20 11 6 10 10 14 8 9 9 12 9 (to the nearest pound).

This gives $\bar{x} = 10.6$.

Rewriting the observations in ascending order of magnitude we have

6 8 9 9 9 9 10 10 11 12 14 20.

As there are an even number of observations the median is the average of the sixth and seventh values, namely nine and a half.

As eight of the observations are less than the sample mean, it could be argued that the median is 'more typical' of the data. An outlying observation (in this example, 20) may have a considerable effect on the sample mean but will not have much effect on the median.

The mode. This is the value of the variate which occurs with the greatest frequency. For discrete data the mode can easily be found by inspection. For continuous data the mode can be estimated by plotting the results in a histogram and finding the midpoint of the tallest box. Thus in Example 1 the mode is 67 inches.

Comparison. As we have already remarked, the mean is by far the most important measure of location. When the distribution of results is roughly symmetric, the mean, mode and median will be very close together anyway. But if the distribution is very skewed there may be a considerable difference between them and then it may be useful to find the mode and median as well as the mean. Figure 8 shows the frequency curve for the time taken to repair aircraft faults (R. A. Harvey, 'Applications of statistics to aircraft project evaluation', *Applied Statistics*, 1967). Typical values are one hour for the mode, two hours for the median, and four hours for the mean. Obviously extra care is needed when describing skew distributions.

2.3.2 Measures of spread

It is often equally important to know how spread out the data is. For example suppose that a study of people affected by a certain disease revealed that most people affected were under two years old or over seventy years old; then it would be very misleading to summarize the data by saying 'average age of persons affected is thirty-five years'. At the very least we must add a measure of the variability or spread of the data. Several such quantities are available.

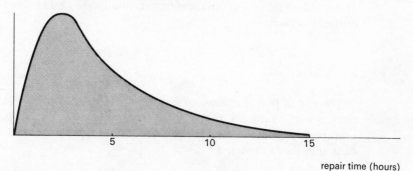

repair time (hours)

Figure 8   Frequency curve of aircraft repair times

Range. This is the difference between the largest and smallest observation. It can be very useful for comparing the variability in samples of equal size but is unfortunately affected by the number of observations; the more observations taken, the larger the range will be. So it is not a fixed characteristic of the population.

Mean absolute deviation or M.A.D. This is the average absolute deviation from the sample mean. For n observations, $x_1, x_2, \ldots, x_n$, we have

$$\text{M.A.D.} = \frac{|x_1 - \bar{x}| + |x_2 - \bar{x}| + \ldots + |x_n - \bar{x}|}{n}$$

$$= \sum_{i=1}^{n} \frac{|x_i - \bar{x}|}{n}.$$

This statistic is easy to understand and is often used by non-statisticians. However, later in the book we will see that there are strong theoretical and practical considerations for preferring a statistic called the *variance*, or its square root which is called the *standard deviation*.

Variance and standard deviation. The sample variance s^2 of n observations, $x_1, x_2, \ldots, x_n$, is given by

$$s^2 = \frac{(x_1 - \bar{x})^2 + (x_2 - \bar{x})^2 + \ldots + (x_n - \bar{x})^2}{(n-1)}$$

$$= \sum_{i=1}^{n} \frac{(x_i - \bar{x})^2}{n-1}. \qquad \textbf{2.2}$$

The standard deviation s of the sample is obtained by taking the square root of the variance.

$$s = \sqrt{\left[\sum_{i=1}^{n} \frac{(x_i - \bar{x})^2}{n-1} \right]}.$$

The use of $(n-1)$ instead of n in the denominator of this formula puzzles many students. On the rare occasions when the true population mean μ is known, the formula for s^2 would indeed have n in the denominator as expected.

$$s^2 = \sum_{i=1}^{n} \frac{(x_i - \mu)^2}{n}.$$

However, since μ is not generally known, the sample mean $\bar{x}$ is used instead. Then it can be shown theoretically that it is better to change the denominator from n to $(n-1)$. (This point is discussed in Section 6.5, where s^2 is found to be an unbiased estimate of the true population variance, σ^2.) It is worth pointing out that the difference between n and $(n-1)$ is only important for small samples.

The standard deviation is in the same units as the original measurements and for this reason it is preferred to the variance as a descriptive measure. However it is often easier from a theoretical and computational point of view to work with variances. Thus the two measures are complementary.

In order to calculate a variance on a desk calculating machine, it is usually more convenient to rearrange **2.2** in the form

$$s^2 = \frac{\left\{ \left(\sum_{i=1}^{n} x_i^2 \right) - n\bar{x}^2 \right\}}{(n-1)}. \qquad \textbf{2.3}$$

Example 8

Find the range, variance and standard deviation of the following 6 observations:

0·9 1·3 1·4 1·2 0·8 1·0

Range $= 1·4 - 0·8 = 0·6$.

$$\bar{x} = \frac{6·6}{6} = 1·1.$$

$$\sum x^2 = 7·54.$$

$$s^2 = \frac{(7·54 - 7·26)}{5} = 0·056,$$

$$s = 0·237.$$

Coefficient of variation. We have seen that the standard deviation is expressed in the same units as the individual measurements. For some purposes it is much more useful to measure the spread in relative terms by dividing the standard deviation by the sample mean. The ratio is called the *coefficient of variation.*

$$\text{(coefficient of variation)} = \frac{s}{\bar{x}}.$$

For example a standard deviation of 10 may be insignificant if the average observation is around 10,000 but may be substantial if the average observation is around 100.

Another advantage of the coefficient of variation is that it is independent of the units in which the variate is measured, provided that the scales begin at zero. If every observation in a set of data is multiplied by the same constant, the mean and standard deviation will also be multiplied by this constant, so that their ratio will be unaffected. Thus the coefficient of variation of a set of length measurements, for example, would be the same whether measurements were made in centimetres or inches. However this is not true, for example, for the centigrade and Fahrenheit scales of measuring temperature where the scales do not begin at zero.

Example 9

Using equation **2.3** (or by coding – see exercise 3 below), the standard deviation of the data from Example 1 was found to be

$s = 6\cdot0.$

It is not immediately clear if this is to be judged 'large' or 'small'. However the coefficient of variation is given by

$$\frac{s}{\bar{x}} = \frac{6\cdot0}{1000\cdot6} = 0\cdot006$$

and from this we judge that the variation in the data is relatively small.

Exercises

1. The weight of an object was measured thirty times and the following observations were obtained.

6·120	6·129	6·116	6·114	6·112
6·119	6·119	6·121	6·124	6·127
6·113	6·116	6·117	6·126	6·123
6·123	6·122	6·118	6·120	6·120
6·121	6·124	6·114	6·121	6·120
6·116	6·113	6·111	6·123	6·124

All measurements are to the nearest 0·001 g. Plot a histogram of the observations.

2. Find the mean, variance and standard deviation of the following samples.

(a) 5 2 3 5 8
(b) 2 20 5 3 5

3. The computation of the mean and variance of a set of data can often be simplified by subtracting a suitable constant from each observation. This process is called 'coding'. Find the mean of the coded observations and then add the constant to get the sample mean. The variance (and standard deviation) of the observations is not affected by coding.

Calculate $\bar{x}$ and s^2 for the following samples.

(a) 997 995 998 992 995 (here 990 is a suitable constant)
(b) 131·1 137·2 136·4 133·2 139·1 140·0

Part two
Theory

'Mutton first, mechanics afterwards.'

Chapter 3
The concept of probability

3.1 Probability and statistics

We begin our study of the theory of statistics with an introduction to the concept of probability. Most of us have a good grasp of probability as illustrated by the popularity of games of chance such as bridge, poker, roulette, and dice, all of which involve the assessment of probability. It turns out that the theory of probability is a good base for the study of statistics – in fact each is the inverse subject of the other. The relation between probability and statistics may be clarified with an example.

A *sample* of size one hundred is taken from a very large *batch* (or *lot*) of valves, which contain a proportion p of defective items. This is another situation in which experimental uncertainty is involved, because, if several such samples are taken, the number of defective items will not be the same in each sample.

Let us hypothetically suppose that we know that $p = 0 \cdot 1$. Then we would expect to get $(100 \times 0 \cdot 1) = 10$ defectives in the sample. But a single sample may contain any number of defectives 'close' to 10. If we keep taking samples size 100 from the batch, then sometimes 8 defectives will be present, sometimes 11, sometimes 9, 10, 12 etc. *Probability theory* enables us to calculate the chance or probability of getting a given number of defectives, and we shall see how this is done in Chapter 4.

However, in a typical experimental situation we will not know what p is. Instead, suppose we take a sample size 100 and get 10 defectives. Then what is p? The obvious answer is $10/100 = 0 \cdot 1$; but it may be $0 \cdot 11$ or $0 \cdot 09$ or any number 'close' to $0 \cdot 1$. *Statistical theory* enables us to *estimate* the value of p.

The duality between probability and statistics can be expressed more generally in the following way. In both subjects the first step in solving a problem is to set up a mathematical model for the physical situation which may involve one or more parameters. In the above example the proportion of defectives p is a parameter. If the parameters of the model are known, given or established from past history then we have a probability problem and we can deduce the behaviour of the system from the model. However if the parameters are unknown and have to be estimated from the available data then we have a statistical problem. But to understand and solve statistical problems, it is necessary to have some prior knowledge of probability, and so we will devote the next three chapters to a study of probability and related topics.

Example 1

An air-to-air missile has a 'kill' ratio of 2 in 10. If 4 missiles are launched, what is the probability that the target will not be destroyed? Is this a statistics or probability problem?

The physical situation in this problem can be described by one quantity: the 'kill' ratio. As this is known from previous tests we can calculate the probability that a given number of missiles will hit the target. Thus we have a probability problem. The solution is obtained by considering the 'miss' ratio which must be 8 in 10. The probability that all 4 missiles will miss the target, assuming that they are independently launched, is obviously less than the probability that just one missile will miss the target. Later in the chapter we will see that the probability is given by

$(0.80)^4 = 0.4096.$

Example 2

A hundred missiles are launched and eleven 'kills' are observed. What is the best estimate of the 'kill' ratio?

As the 'kill' ratio was unknown before the test was performed, we have to estimate this quantity from the observations. The ratio of kills to launches is the intuitive estimate, namely 0·11. This is a statistical problem.

3.2 Some definitions

We begin our study of probability theory with some definitions.

The *sample space* is defined as the set of all possible outcomes of an experiment. For example:

When a die is thrown the sample space is 1, 2, 3, 4, 5 and 6.

If two coins are tossed, the sample space is head–head, tail–tail, head–tail, tail–head.

In testing the reliability of a machine, the sample space is 'success' and 'failure'.

Each possible outcome is a *sample point*. A collection of sample points with a common property is called an *event*.

If a die is thrown and a number less than 4 is obtained, this is an event containing the sample points 1, 2 and 3.

If two coins are tossed and at least one head is obtained, this is an event containing the sample points head–head, tail–head, and head–tail.

If the number of particles emitted by a radioactive source in one minute is measured, the sample space consists of zero and all the positive integers, and so is infinite. Obtaining less than five particles is an event which consists of the sample points 0, 1, 2, 3 and 4.

The *probability* of a sample point is the proportion of occurrences of the sample point in a long series of experiments. We will denote the probability that sample point x will occur by $P(x)$. For example, a coin is said to be 'fair' if heads and tails are equally likely to occur, so that $P(\text{H}) = P(\text{T}) = \frac{1}{2}$. By this we mean that if the coin is tossed N times and f_H heads are observed, then the ratio f_H/N tends to get closer to $\frac{1}{2}$ as N increases. On the other hand if the coin is 'loaded' then the ratio f_H/N will not tend to $\frac{1}{2}$.

The probability of a sample point always lies between zero and one. If the sample point cannot occur, then its probability is zero, but if the sample point must occur, then its probability is one. (Note that the converse of these two statements is not necessarily true.) Furthermore the sum of all the probabilities of all the sample points is one.

Probability theory is concerned with setting up a list of rules for manipulating these probabilities and for calculating the probabilities of more complex events.

Example 3

Toss two fair coins. Denote heads by H and tails by T. There are four points in the sample space and they are equally likely to occur.

Sample space	HH	HT	TH	TT
Probability	$\frac{1}{4}$	$\frac{1}{4}$	$\frac{1}{4}$	$\frac{1}{4}$

Example 4

Throw a fair die.

Sample space	1	2	3	4	5	6
Probability	$\frac{1}{6}$	$\frac{1}{6}$	$\frac{1}{6}$	$\frac{1}{6}$	$\frac{1}{6}$	$\frac{1}{6}$

The probability of an event is the sum of the probabilities of the sample points which constitute the event.

Example 5

If two fair coins are tossed, obtaining at least one head is an event, denoted by E, which consists of the three sample points (HH), (HT), and (TH).

Probability of this event $= P(E)$
$$= P(HH) + P(HT) + P(TH)$$
$$= \tfrac{1}{4} + \tfrac{1}{4} + \tfrac{1}{4} = \tfrac{3}{4}.$$

Example 6

If two fair dice are tossed, the sample space consists of the thirty-six combinations shown below.

$$
\begin{array}{cccccc}
1,1 & 1,2 & 1,3 & 1,4 & 1,5 & 1,6 \\
2,1 & 2,2 & 2,3 & 2,4 & 2,5 & 2,6 \\
3,1 & 3,2 & 3,3 & 3,4 & 3,5 & 3,6 \\
4,1 & 4,2 & 4,3 & 4,4 & 4,5 & 4,6 \\
5,1 & 5,2 & 5,3 & 5,4 & 5,5 & 5,6 \\
6,1 & 6,2 & 6,3 & 6,4 & 6,5 & 6,6
\end{array}
$$

Each of the thirty-six sample points is equally likely to occur and so each has probability $\frac{1}{36}$. By inspection we can see for example that

$P(\text{sum of the 2 dice is 7}) = \frac{6}{36} = \frac{1}{6}$

$P(\text{sum is 2}) = \frac{1}{36}$

$P(\text{sum is 7 or 11}) = \frac{6}{36} + \frac{2}{36} = \frac{8}{36}.$

3.3 Types of events

3.3.1 *Mutually exclusive events*

If two events, E_1 and E_2, are mutually exclusive, they have no common sample points. In a single trial mutually exclusive events cannot both occur; the probability that one of the mutually exclusive events occurs being the sum of their respective probabilities. This is the *addition law* for mutually exclusive events.

$$P(E_1 + E_2) = P(E_1) + P(E_2).$$ **3.1**

The notation $(E_1 + E_2)$ means that at least one of the events occurs; as applied to mutually exclusive events it means that one event or the other occurs.

Example 7

From a fair pack of well-shuffled cards one card is dealt. Define the event E_1 as 'the card is a spade' and event E_2 as 'the card is a heart'. These two events are mutually exclusive as one card cannot be both a heart and a spade.

Then $P(E_1 + E_2) = \frac{13}{52} + \frac{13}{52} = \frac{1}{2}$.

3.3.2 *Not mutually exclusive*

Two events that are not mutually exclusive contain one or more common sample points. The probability that at least one of these events occur is given by the *general addition law*

$$P(E_1 + E_2) = P(E_1) + P(E_2) - P(E_1 E_2),$$ **3.2**

where $(E_1 E_2)$ is the event that *both* E_1 and E_2 occur.

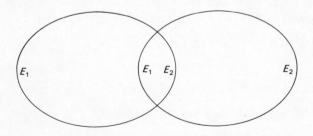

Figure 9   Events E_1 and E_2 are not mutually exclusive

The general addition law may be demonstrated pictorially. Let the area of a unit square represent the probability of the entire sample space. Inside this draw two figures with areas equal to the respective probabilities of E_1 and E_2 and such that the common area is equal to the probability that both the events will occur.

The area of the event $(E_1 + E_2)$ is the sum of the areas of E_1 and E_2 minus their common area $E_1 E_2$, which is included in both E_1 and E_2. Thus we have

$$P(E_1 + E_2) = P(E_1) + P(E_2) - P(E_1 E_2).$$

Note that if E_1 and E_2 are mutually exclusive, the probability of the joint event $(E_1 E_2)$ is zero – that is, $P(E_1 E_2) = 0$ – and then equation **3.2** reduces to equation **3.1**. Mutually exclusive events have no common area and can be depicted as in Figure 10.

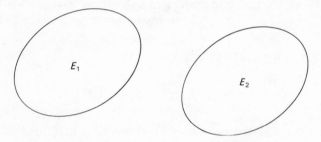

Figure 10   Events E_1 and E_2 are mutually exclusive

Example 8

What is the probability of rolling two dice to obtain the sum seven and/or the number three on at least one die? Let event E_1 be 'the sum is 7' and event E_2 be 'at least one 3 turns up'. By inspection of the table in Example 6, we find

$$P(E_1) = \tfrac{6}{36} \quad P(E_2) = \tfrac{11}{36}$$

$$P(E_1 E_2) = \tfrac{2}{36}.$$

Thus $P(E_1 + E_2) = P(E_1) + P(E_2) - P(E_1 E_2) = \tfrac{6}{36} + \tfrac{11}{36} - \tfrac{2}{36} = \tfrac{15}{36}.$

This can be checked by looking at the table.

Events that are not mutually exclusive may be further classified as *dependent* or *independent* events. Dependence between events is treated by the notion of *conditional probability*.

3.3.3 Conditional probability

We defined the probability of an event as the sum of the probabilities of the sample points in the event. $P(E) = \sum_{\text{event}} P(\text{sample points in E})$. In a more general sense we could have defined the probability of an event as follows:

$$P(E) = \frac{\sum\limits_{\text{event}} P(\text{sample points in E})}{\sum\limits_{\substack{\text{sample} \\ \text{space}}} P(\text{all sample points})}.$$ **3.3**

The denominator is of course unity.

Now suppose that we are interested in the probability of an event E_1 and we are told that event E_2 has occurred. The conditional probability of E_1, given that E_2 has occurred, is written $P(E_1|E_2)$, and read as $P(E_1$ given $E_2)$. By analogy with equation **3.3**, this conditional probability can be defined:

$$P(E_1|E_2) = \frac{\sum\limits_{E_1 E_2} P(\text{sample points common to } E_1 \text{ and } E_2)}{\sum\limits_{E_2} P(\text{sample points in } E_2)}$$

$$= \frac{P(E_1 E_2)}{P(E_2)}.$$ **3.4**

The effect of the conditional information is to restrict the sample space to the sample points contained in event E_2.

Example 9

Given that a roll of two fair dice has produced at least one three, what is the probability that the sum is seven? Event E_1 is 'the sum is 7' and event E_2 is 'at least one 3 turns up'.

$$P(E_1|E_2) = \frac{P(E_1 E_2)}{P(E_2)}$$

$$= \frac{2/36}{11/36} \quad \text{(see Example 8)}$$

$$= \tfrac{2}{11}.$$

This result can be obtained directly from the table in Example 6, by removing all points in which no three occurs. Of the remaining eleven points exactly two give a sum which is seven.

Example 10

Given that the last roll of two dice produced the sum seven, what is the probability that the next roll will also produce the sum seven? Careful! If the dice are fair dice and the roll is preceded by a good shake in a dice cup, the play is *fair*, which implies that dice have no memory. Thus there is no connexion between the results of successive rolls, and they are independent events rather than dependent events. In other words, the knowledge that the last roll was a seven contributes nothing to our ability to predict the next roll.

3.3.4 *Independent and dependent events*

Two events, E_1 and E_2, are said to be *independent* if $P(E_1) = P(E_1|E_2)$. Thus the knowledge that event E_2 has occurred has no effect on the probability of event E_1.

Conversely, two events are said to be *dependent* if $P(E_1) \neq P(E_1|E_2)$.

Example 10 (continued)

As two successive rolls are independent the probability of the sum seven on the next roll is still $\frac{6}{36}$. The last result is irrelevant. This is a subtle point.

Some people interpret the 'law of averages' as meaning that if the last roll was a seven the next roll is less likely to be a seven; that the dice will act to 'even out' the results. The implication is that the dice have some kind of memory.

A diametrically opposed motion is that if sevens have been 'popular' they will continue to be. This idea is based on the belief that most *real* gambling equipment will be slightly biased and therefore some events will occur more often than expected. For this reason honest gambling casinos make extreme efforts to keep their games of chance fair. It is in their best interest. Most serious studies of casino data discredit both the above theories. Games of chance do not have memories.

3.3.5 *Joint events*

The probability of the joint event $(E_1 E_2)$ can be obtained by considering

$$P(E_1|E_2) = \frac{P(E_1 E_2)}{P(E_2)},$$

which can be rearranged to give

$P(E_1 E_2) = P(E_2)P(E_1|E_2)$.

Similarly $P(E_1 E_2) = P(E_1)P(E_2|E_1)$.

These relations are general and apply to both dependent and independent events. Of course if the events are independent then $P(E_1|E_2) = P(E_1)$, so that the equation simplifies to give

$P(E_1 E_2) = P(E_1)P(E_2)$.

This is called the *product law for independent events*.

Example 11

What is the probability of rolling the sum seven with two fair dice, one of which will be a three? Event E_1 is 'the sum is 7', event E_2 is 'at least one 3 occurs'.

$$P(E_1 E_2) = P(E_2)P(E_1|E_2)$$
$$= \tfrac{11}{36} \times \tfrac{2}{11} \quad \text{(see Example 9)}$$
$$= \tfrac{1}{18}.$$

Example 12

What is the probability of drawing two aces from a well-shuffled pack?

Event $E_1 =$ ace on the first card

Event $E_2 =$ ace on the second card

$P(E_2|E_1)$ is obtained by noting that if the first card drawn is an ace, then there will be fifty-one cards left in the pack of which three are aces.

Thus $P(E_1 E_2) = P(E_1)P(E_2|E_1)$
$$= \tfrac{4}{52} \times \tfrac{3}{51}$$
$$= 0\cdot0045.$$

Although two cards are drawn in the above experiment, the student may think of the situation as a single trial. We can write

$E_1 =$ ace on first card together with any second card,
$E_2 =$ ace on second card together with any first card,
$E_1 E_2 =$ ace on first and second cards.

Alternatively the student may think of the experiment as two trials. The result of the second trial (the second card) depends on the result of the first trial (the first card). The laws of probability give the same result whichever way the problem is studied.

Example 13

A fair die is thrown. What is the probability that it will show 6 on the first throw and 2 on the second throw?

$E_1 = 6$ on first throw,
$E_2 = 2$ on second throw.

But the occurrence or non-occurrence of event E_1 has no bearing on whether or not E_2 occurs. Thus the two events are independent.

$$P(E_1 E_2) = P(E_1)P(E_2)$$
$$= \tfrac{1}{6} \times \tfrac{1}{6}$$
$$= \tfrac{1}{36}.$$

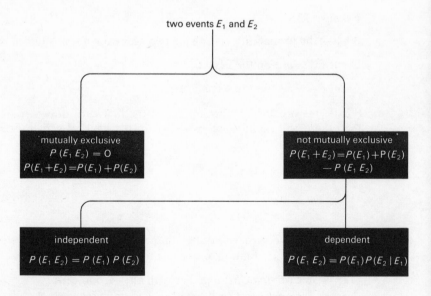

Figure 11   Types of events

3.3.6 Summary of types of events

At first reading the classification of events may be a little confusing, but Figure 11 may help to clarify the situation.

Example 14

If there are three events then similar reasoning and diagrams will provide all the required relationships.

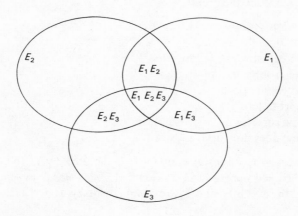

Figure 12 Three events

By inspection,

$$P(E_1 + E_2 + E_3) = P(E_1) + P(E_2) + P(E_3) - P(E_1 E_2) - P(E_1 E_3)$$
$$- P(E_2 E_3) + P(E_1 E_2 E_3),$$

$$P(E_1|E_2 E_3) = \frac{P(E_1 E_2 E_3)}{P(E_2 E_3)},$$

$$P(E_1 E_2 E_3) = P(E_1)P(E_2|E_1)P(E_3|E_1 E_2).$$

If E_1, E_2 and E_3 are independent events then

$$P(E_1 E_2 E_3) = P(E_1)P(E_2)P(E_3).$$

Example 15

A manned rocket vehicle is estimated to have a reliability (probability of mission success) of 0·81 on the first launch. If the vehicle fails there is a probability of 0·05 of a catastrophic explosion, in which case the abort system cannot be used. If an abort can be attempted, the abort system reliability is 0·90.

Calculate the probability of every possible outcome of the first launch.

First step. Define the events.

E_1 – Mission success
$\overline{E}_1$ – Mission failure
E_2 – Non-catastrophic failure
$\overline{E}_2$ – Catastrophic failure
E_3 – Successful abort
$\overline{E}_3$ – Abort failure
E_4 – Crew survives
$\overline{E}_4$ – Crew does not survive.

In each case the opposite event of E is denoted by $\overline{E}$. (Note that if $P(E) + P(\overline{E}) = 1$ the events E, $\overline{E}$ are called *complementary* events. In this example, the events E_1 and $\overline{E}_1$ are complementary, as well as E_4 and $\overline{E}_4$.)

Second step. Classify the events. For example a mission cannot be both a success and a non-catastrophic failure, so that E_1 and E_2 are mutually exclusive. In fact E_1 is mutually exclusive with $\overline{E}_2$, E_3, $\overline{E}_3$ and $\overline{E}_4$. E_2 is mutually exclusive with E_3, $\overline{E}_3$ and E_4. E_3 is mutually exclusive with $\overline{E}_4$. Also each event is mutually exclusive with its opposite event.

Third step. The probabilities of the events can be found in succession.

$$P(E_1) = 0·81,$$
$$P(\overline{E}_1) = 1 - P(E_1)$$
$$= 0·19.$$

$P(\overline{E}_1 \overline{E}_2) = P(\overline{E}_2|\overline{E}_1)P(\overline{E}_1)$ from equation **3.4**. But a catastrophic failure ($\overline{E}_2$) can only occur if a mission failure ($\overline{E}_1$) occurs. Thus we have

$$\overline{E}_1 \overline{E}_2 = \overline{E}_2.$$

Therefore $P(\overline{E}_2) = P(\overline{E}_2|\overline{E}_1)P(\overline{E}_1)$

$$= 0\cdot05 \times 0\cdot19$$

$$= 0\cdot0095.$$

By similar reasoning we have

$P(E_2) = P(E_2\overline{E}_1)$

$$= P(E_2|\overline{E}_1)P(\overline{E}_1)$$

$$= 0\cdot95 \times 0\cdot19$$

$$= 0\cdot1805.$$

$P(E_3) = P(E_3E_2)$

$$= P(E_3|E_2)P(E_2)$$

$$= 0\cdot90 \times 0\cdot1805$$

$$= 0\cdot16245.$$

$P(E_4) = P(E_1) + P(E_3)$

$$= 0\cdot81 + 0\cdot16245$$

$$= 0\cdot97245$$

$P(\overline{E}_4) = 1 - P(E_4)$

$$= 0\cdot02755.$$

$P(\overline{E}_3) = P(\overline{E}_3E_2)$

$$= P(\overline{E}_3|E_2)P(E_2)$$

$$= 0\cdot10 \times 0\cdot1805$$

$$= 0\cdot01805.$$

Check: $P(\overline{E}_4) = P(\overline{E}_2) + P(\overline{E}_3)$

$$0\cdot02755 = 0\cdot0095 + 0\cdot01805.$$

3.4 Permutations and combinations

Many problems in probability require the total number of points in a sample space to be counted. When there is a very large number of these, a knowledge of combinatorial theory, and in particular of permutations and combinations, is useful.

A quantity frequently used in combinatorial theory is *factorial n*, which is defined as factorial $n = n! = n(n-1)(n-2)\ldots 3 \times 2 \times 1$.

3.4.1 *Permutations*

The number of ways in which r items can be selected from n distinct items, taking notice of the order of selection, is called the number of permutations of n items taken r at a time, and is denoted by $^{n}P_{r}$, or $P(n, r)$.

$$^{n}P_{r} = n(n-1)\ldots(n-r+1)$$

$$= \frac{n!}{(n-r)!}.$$

For example, the number of permutations of the letters a, b, c taken two at a time is

$$^{3}P_{2} = 3 \times 2 = 6.$$

These are ab, ac, bc, ba, ca, cb.

3.4.2 *Combinations*

The number of ways in which r items can be selected from n distinct items, disregarding the order of the selection, is called the number of combinations of n items taken r at a time. This is denoted by $^{n}C_{r}$, or $\binom{n}{r}$.

There are $r!$ permutations which are the same combination. Thus we have

$$^{n}C_{r} = \frac{^{n}P_{r}}{r!} = \frac{n!}{(n-r)!\,r!} = {}^{n}C_{n-r}.$$

For example, the number of combinations of the letters a, b, c taken two at a time is

$$^{3}C_{2} = \frac{6}{2!} = 3.$$

These are ab, ac, bc.

Important note: ab is the same combination as ba but *not* the same permutation.

The above expressions enable us to solve many problems in probability.

Example 16

Consider the probability of being dealt a bridge hand which consists of all thirteen cards of one suit.

The total number of possible hands is the number of combinations of thirteen cards out of fifty-two. (We are not concerned with the order in which the cards are dealt.) This is $^{52}C_{13}$. The total number of perfect hands is just four (all hearts, all spades, all diamonds or all clubs). If all possible hands are equally likely then

$$\text{probability (perfect hand)} = \frac{4}{^{52}C_{13}}.$$

This is very small indeed ($<10^{-10}$). Most such events which are recorded in the newspapers are probably hoaxes.

Example 17

An inspector draws a sample of five items from a batch of a hundred valves which are numbered from one to a hundred to distinguish them. How many distinct samples can he choose? If there is one defective item in the batch, how may distinct samples of size five can be drawn which contain the defective item? What is the probability of drawing a sample which contains the defective item?

As we are not concerned with the order in which the valves are selected, the total number of distinct samples is the number of combinations of five from a hundred, that is, $^{100}C_5$. In the second case one of the items is fixed and we need to find the total number of ways of selecting four more items from the remaining ninety-nine valves. This is $^{99}C_4$.

Thus the probability of drawing a sample which contains the defective item is

$$\frac{^{99}C_4}{^{100}C_5} = \frac{5}{100}.$$

This probability can of course be written down straight away by noting that any valve has a probability of $\frac{5}{100}$ of being in any particular sample.

3.5 Random variables

One of the basic ideas in probability is that of a random variable. In many cases this is simply the numerical variable under consideration.

For example, if a penny is tossed twice, the number of heads which turn up can be either 0, 1 or 2 according to the outcome of the experiment. We say that the number of heads is a random variable, since it expresses the result of the experiment as a number. Other simple random variables are the temperature of a chemical process and the resistance of a thermocouple.

More generally a random variable is a rule for assigning a particular number to a particular experimental outcome. For example, if the result of an experiment is recorded as say 'the experiment was successful' or 'the experiment was not successful', then to obtain a random variable we must code the results so that, for example, a success corresponds to 1 and a failure to 0. We can also consider more complicated random variables. For example, if a penny is tossed twice, the random variable $X = $ (number of heads)^2 can take the values 0, 1 or 4 according to the outcome of the experiment. The sample variance is another example of a random variable, since for any given series of observations, $x_1, x_2, \ldots, x_n$, there is a corresponding number

$$s^2 = \sum_{i=1}^{n} \frac{(x_i - \bar{x})^2}{(n-1)},$$

which varies from sample to sample.

Mathematically a random variable is a numerically valued function defined on the sample space. This means that for every possible experimental outcome (sample point), there is a corresponding number. This number is one way of expressing the information which results from the experiment.

We have already discussed in section 2.1 the difference between discrete and continuous experimental data. Similarly a random variable is said to be discrete if it can only take a discrete set of values, and is said to be continuous if it can take any value in some specified range.

Example 18

The items in a sample of ten electrical units are classified as defective or non-defective. Is the number of defectives in the sample, X, a discrete or continuous random variable?

The number of defectives in the sample can be any integer between 0 and 10. Thus X is a discrete random variable (it cannot take the value $1\frac{1}{2}$ for example).

Example 19

Consider the random variable X = length of a screw. This variable can be any positive number so that it is a continuous variable. However if the length is *measured*, then the accuracy of the measurements is limited by the accuracy of the measuring instrument so that in practice the measured length can only take discrete values. Thus the distinction between discrete and continuous data is not clear-cut. If the length is measured to two decimal places and we observe say 2·40 inches, then

probability (measured length = 2·40 inches)

$\qquad$ = probability $(2\cdot395 < X < 2\cdot405)$.

Although all the probability concepts discussed this far have been illustrated with discrete examples, the ideas are equally applicable to continuous variables, with one important distinction. The number of sample points in a continuous sample space is always infinite and the probability of a single sample point is zero. As in Example 19 the probability of a single sample point is replaced by the idea of the probability of an interval, which contains an infinite number of sample points.

In Chapter 4 we will continue our discussion of discrete random variables and will return to continuous variables in Chapter 5.

Exercises

1. Two parts of a machine are manufactured independently. These two parts have probabilities p_1 and p_2 respectively of failing. Find the probability that
(a) neither part fails
(b) at least one part fails
(c) exactly one part fails.
2. A component A which has 90 per cent reliability (it will fail one in ten times) is vital to the running of a machine. An identical component is fitted in parallel to form system S_1, and the machine will work provided that one of these components functions correctly.

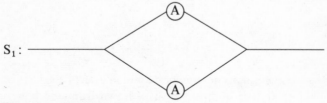

S_1:

Find the reliability of system S_1.

3. In two rolls of a fair die what is the probability that
(a) the two outcomes are the same?
(b) the second outcome is larger than the first?
(c) the second outcome is a five?

4. What is the probability of getting
(a) at least three heads in four tosses of a fair coin?
(b) at least two tails in four tosses of a fair coin? (Assume that all $2^4 = 16$ outcomes are equally likely.)

5. A ball is chosen at random out of an urn containing three black balls and two red balls and then a selection is made from the remaining four balls. Find the probability that a black ball will be selected
(a) at the first time
(b) the second time
(c) both times (assume all twenty outcomes are equally likely).

6. How many possible pairs consisting of a president and a vice-president can be formed from a club of sixty members, provided no member can hold both offices?

7. Special alloys are usually produced in batches called 'melts'. Any castings, forgings or machined parts which are made from these alloys carry an identification of the melt because there may be significant variation from melt to melt. The usual identification employed is three letters. How many melt designations are possible?

8. A group of randomly selected people compare the months and days of their birthdays. How large a group is required to have at least an even chance of finding two people with the same birthday? (Ignore leap years.) Hint: Define the event E_2 as the second birthday does not match the first; the event E_3 as the third does not match the first or second; and so on. Then P(No match between n people)

$$= P(E_n)$$

$$= P(E_2)P(E_3|E_2)P(E_4|E_3 E_2) \ldots P(E_n|E_{n-1} \ldots E_2).$$

9. If a batch contains ninety good and ten defective valves, what is the probability that a sample size five drawn from the batch will not contain any defective items?

10. (a) In order that the system

S_1:

consisting of components A and B should function correctly, both components must function correctly. Assuming that each component functions independently of the other, find the probability that S_1 functions, given that A, B have probabilities 0·8, 0·9 respectively of functioning correctly.

(b) The system

S_2: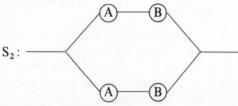

is connected in parallel in such a way that if either of the sub-systems A–B functions correctly, S_2 functions correctly. Find the probability that S_2 functions correctly.

(c) The system

S_3: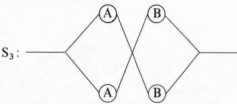

functions correctly if either A and either B function correctly. Find the probability that S_3 functions correctly.

11. You are playing Russian roulette with a gun which has six chambers. What is the probability of someone being killed during the first six pulls of the trigger? (The chamber is spun after each firing and you may assume that the probability of hitting a loaded chamber $= \frac{1}{6}$.)

Chapter 4
Discrete distributions

4.1 The discrete probability distribution

One of the most important concepts in probability is the idea of a probability distribution. This chapter deals with discrete distributions, while continuous distributions are considered in Chapter 5. We begin with a formal definition. If a discrete random variable, X, can take values $x_1, x_2, \ldots, x_N$, with probabilities $p_1, p_2, \ldots, p_N$, where $p_1 + p_2 + \ldots + p_N = 1$ and $p_i \geqslant 0$ for all i, then this defines a *discrete probability distribution* for X.

The probability that X takes a particular value x will be denoted by $P(X = x)$ or simply $P(x)$.

Example 1

Toss two fair coins. Consider the random variable X = number of heads which turn up. Then X can take the values 0, 1 or 2 according to the outcome of the experiment. From Example 3, Chapter 3 we know $P(X = 0) = \frac{1}{4}$, $P(X = 1) = \frac{1}{2}$, $P(X = 2) = \frac{1}{4}$. The sum of these probabilities is one. Thus this random variable follows a discrete probability distribution.

Example 2

Throw two fair dice. Consider the random variable X = (sum of the pips on the two dice). Then X can take the values $2, 3, 4, \ldots, 11, 12$ according to the outcome of the experiment. By inspection of the table in Example 6, Chapter 3, we find

$P(2) = \frac{1}{36}$ $P(3) = \frac{2}{36}$ $P(4) = \frac{3}{36}$ $P(5) = \frac{4}{36}$ $P(6) = \frac{5}{36}$

$P(7) = \frac{6}{36}$ $P(8) = \frac{5}{36}$ $P(9) = \frac{4}{36}$ $P(10) = \frac{3}{36}$ $P(11) = \frac{2}{36}$

$P(12) = \frac{1}{36}$.

These probabilities sum to unity and we have another discrete distribution.

The discrete probability distribution is a very important tool in probability and statistics and in the remainder of the chapter we will consider two valuable distributions which frequently arise in practice.

4.2 The binomial distribution

This distribution has a wide range of practical applications, ranging from sampling inspection to the failure of rocket engines.

Suppose that a series of n independent trials is made, each of which can be a success, with probability p, or a failure, with probability $(1 - p)$. The number of successes which is observed may be any integer between 0 and n. In section 4.3 we will show that

(probability of getting r successes out of n) = $P(r)$

$$= {}^nC_r p^r (1-p)^{n-r} \qquad (r = 0, 1, \ldots, n).$$

These probabilities define a discrete probability distribution customarily called the *binomial distribution*. The probabilities are obtained when the expression

$$1 = [p + (1-p)]^n = \sum_{r=0}^{n} {}^nC_r p^r (1-p)^{n-r}$$

is expanded by the binomial theorem. In fact this is probably the easiest way to remember the binomial probabilities. It also explains why the distribution is called the binomial distribution and proves that the sum of the probabilities is one.

A discrete distribution can be represented pictorially by a bar chart. Figure 13 show four examples of the binomial distribution, with different values for n and p, to show the various distribution shapes which can occur.

Figure 13(c) is the mirror image of Figure 13(b), since the values for n are the same and $0.9 = 1 - 0.1$. When $p = 0.5$ the binomial distribution is symmetric about the point $\frac{1}{2}n$; Figure 13(d) is an example of this with $n = 20$.

The binomial distribution has been tabulated for $n = 1(1)49$ and $p = 0.01(0.01)0.50$ in National Bureau of Standards (1950). For larger values of n it becomes very tedious to compute the binomial probabilities and then it is often convenient to approximate the binomial distribution either with the Poisson distribution (see section 4.6) or with the normal distribution (see section 5.4).

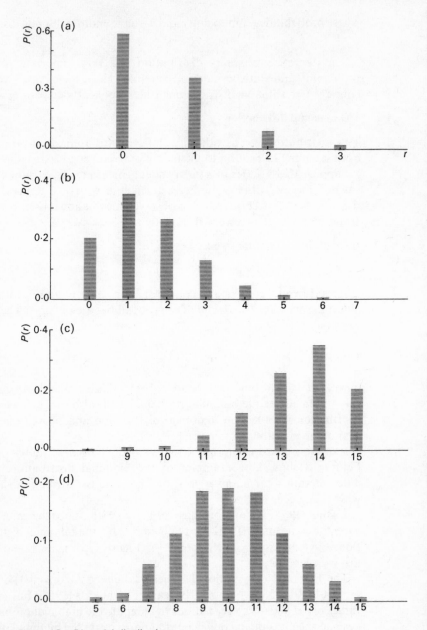

Figure 13   Binomial distributions
(a) reverse J-shaped, $n = 3$, $p = \frac{1}{3}$ (b) positively skewed or skewed to the right, $n = 15$, $p = 0.1$ (c) negatively skewed or skewed to the left, $n = 15$, $p = 0.9$ (d) symmetric, $n = 20$, $p = 0.5$

4.3 The binomial model

The binomial distribution is applicable whenever a series of trials is made which satisfies the following conditions.

(1) Each trial has only two possible outcomes, which are mutually exclusive. We will usually call the two outcomes success (denoted by S) and failure (denoted by F). Other possible pairs of outcomes are defective or non-defective, go or no-go, heads or tails.

(2) The probability of a 'success' in each trial is a constant, usually denoted by p. Thus the probability of a 'failure' is $(1-p)$.

(3) The outcomes of successive trials are mutually independent.

A typical situation in which these conditions will apply (at least approximately) occurs when several items are selected at random from a very large batch and examined to see if there are any defective items (that is, failures). The number of defectives in a sample size n is a random variable, denoted by X, which can take any integer value between 0 and n. In order to find the probability distribution of X, we first consider the simple case when $n = 2$.

4.3.1 *2 trials*

If two trials are made, the sample space consists of the four points SS, SF, FS and FF.

$$\text{Probability (2 successes)} = P(\text{SS})$$

$$= P(\text{S})P(\text{S}) \quad \text{by the assumption of independence}$$

$$= p^2.$$

$$\text{Probability (1 success)} = P(\text{SF}) + P(\text{FS})$$

$$= p(1-p) + (1-p)p.$$

$$\text{Probability (0 successes)} = P(\text{FF})$$

$$= (1-p)^2.$$

Thus the probability distribution of the number of successes in two trials is

$$P(0) = (1-p)^2 \qquad P(1) = 2p(1-p) \qquad P(2) = p^2.$$

Note that $P(0) + P(1) + P(2) = 1$ as required.

This distribution can be obtained by substituting $n = 2$ into the general formula for the binomial distribution which was given in the previous section.

Example 3

Toss two fair pennies. Let X = number of heads showing. If we think of a head as a success and a tail as a failure we have

p = (probability of a head in one throw)

$= \frac{1}{2}$.

Substituting this value of p into the above formulae we have

$P(0) = \frac{1}{4}$ $P(1) = \frac{1}{2}$ $P(2) = \frac{1}{4}$,

as already obtained in Example 1.

4.3.2   *n trials*

More generally we will consider the case where n trials are carried out. The probability of getting r successes in n trials can be found as follows;

$P(\text{exactly } r \text{ successes}) = P(r \text{ successes and } (n-r) \text{ failures})$

There are many possible sequences which can give exactly r successes. For example, we can start with r successes and finish with $(n-r)$ failures. Since successive trials are independent we have

$$\overset{\longleftarrow r \longrightarrow}{} \overset{\longleftarrow n-r \longrightarrow}{}$$
$$P(SS \ldots S \ FF \ldots F) = p^r(1-p)^{n-r}.$$

Every possible ordering of r successes and $(n-r)$ failures will have the same probability, $p^r(1-p)^{n-r}$. The number of possible orderings is the number of combinations of r out of n. Thus we have

$P(\text{exactly } r \text{ successes}) = {}^nC_r p^r(1-p)^{n-r}$ $(r = 0, 1, 2, \ldots, n)$

and so the number of successes follows the binomial distribution.

In order to derive the binomial distribution in the above way, we have carefully stated certain assumptions about the behaviour of successive trials. These assumptions constitute what is called the binomial model. This model is a typical example of a mathematical model in that it attempts to describe a particular physical situation in mathematical terms. Such models may depend on one or more *parameters* which specify how the system will behave. In the binomial model the quantities n and p are parameters. If the values of the model parameters are known, it is a simple matter to calculate the probability of a particular outcome. We will now illustrate the use of the binomial distribution with some examples.

Example 4

Roll three dice. Find the distribution of the random variable $X =$ (number of twos which turn up).

X can take the values 0, 1, 2 or 3. Now the probability of getting a two for one die is $\frac{1}{6}$. An incorrect but popular deduction is that the probability of getting at least one two from three dice is $\frac{3}{6}$. Let us call the event 'getting a two' a success so that the probability of a success is $\frac{1}{6}$. As the results from the three dice are independent, the binomial model is applicable and the distribution of the number of twos is a binomial distribution with parameters $n = 3$, $p = \frac{1}{6}$.

$$P(0) = {}^3C_0(\tfrac{1}{6})^0(\tfrac{5}{6})^3 = 0\cdot58,$$

$$P(1) = {}^3C_1(\tfrac{1}{6})(\tfrac{5}{6})^2 = 0\cdot35,$$

$$P(2) = {}^3C_2(\tfrac{1}{6})^2(\tfrac{5}{6}) = 0\cdot07,$$

$$P(3) = {}^3C_3(\tfrac{1}{6})^3(\tfrac{5}{6})^0 = 0\cdot005.$$

Note that the probability of getting at least one 2 is 0·42 which is considerably less than $\frac{3}{6}$.

In effect this means that if we roll 3 dice 100 times we would expect to get the following results.

r	Number of occurrences
0	58
1	35
2	7
3	0

Of course if we actually try it we will probably get a slightly different distribution due to sampling fluctuation. The above distribution is what we expect on average.

Example 5

On past evidence, an electrical component has a probability of 0·98 of being satisfactory. What is the probability of getting two or more defectives in a sample size five?

The number of defectives in a sample is a binomial random variable with $p = 0.02$, $n = 5$ so that

$$P(0 \text{ defectives}) = 0.98^5$$

$$= 0.904$$

$$P(1 \text{ defective}) = 5 \times 0.98^4 \times 0.02$$

$$= 0.092.$$

This means that the probability of getting two or more defectives in a sample size five is

$$1 - 0.904 - 0.092 = 0.004$$

which is very small indeed.

In effect this means that if two or more defectives are observed in practice then it is likely that something has gone wrong with the manufacturing process.

Example 6

The latest recoilless rocket launcher has a kill probability of 0.60. How many rockets would you have to fire to have at least a 95 per cent probability of destroying the target? What is the probability distribution of salvos of this size?

Let the required salvo contain n rockets. The probability of success of each rocket is given by $p = 0.60$. Thus the number of rockets which actually hit the target has a binomial distribution with parameters n, and $p = 0.60$. Thus the probability that no rockets hit the target is given by $(1 - p)^n = 0.40^n$.

Hence the probability that the target is destroyed (that is, at least one rocket hits the target) is given by

$$1 - 0.40^n.$$

We are told that this must be greater than 0.95. Therefore,

$$1 - 0.40^n > 0.95$$

$$0.05 > 0.40^n$$

giving $n = 4$ since n must be an integer.

Then the probability distribution for salvos of this size is given by

$$P(r \text{ hits}) = {}^4C_r(0{\cdot}60)^r(0{\cdot}40)^{4-r} \qquad (r = 0, 1, 2, 3, 4)$$

$$P(0 \text{ hits}) = 0{\cdot}026$$

$$P(1 \text{ hit}) = 0{\cdot}154$$

$$P(2 \text{ hits}) = 0{\cdot}346$$

$$P(3 \text{ hits}) = 0{\cdot}346$$

$$P(4 \text{ hits}) = 0{\cdot}130.$$

Example 7

An inspector takes a random sample of ten items from a very large batch. If none of the items is defective he accepts the batch, otherwise he rejects the batch. What is the probability that a batch is accepted if the fraction defective is 0, 0·01, 0·02, 0·05, 0·1, 0·2, 0·5, 1? Plot this probability against the fraction defective.

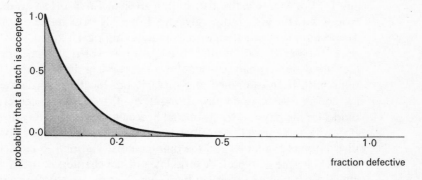

Figure 14

The number of defectives in a sample is a binomial random variable with parameters $n = 10$ and p = fraction defective. Thus

$$P(0 \text{ defectives}) = (1-p)^{10}$$

$$= \text{probability that a batch is accepted.}$$

When $p = 0$ the batch is certain to be accepted; when $p = 1$ the batch is certain to be rejected.

For intermediate values of p the expression $(1-p)^{10}$ must be evaluated. The results are plotted in Figure 14.

4.4 Types of distribution

Before continuing with our discussion of theoretical discrete distributions, we will pause briefly to distinguish between the four different types of distribution which can occur.

Table 5

	Discrete	*Continuous*
Theoretical (parameters known)	type I	type III
Empirical (parameters unknown)	type II	type IV

In Chapter 2 we studied empirical frequency distributions of types II and IV. For example, the distribution given in Table 1 is an example of type II, and the distribution given in Table 2 is an example of type IV. The technologist will often wish to analyse empirical frequency distributions of types II and IV and once again it is worth emphasizing that they are easier to analyse when the corresponding theoretical distributions have been examined and understood. In other words the best method of analysing empirical frequency distributions is to set up a model for the physical situation and hence find the probability distribution which adequately describes the data. There are many theoretical distributions available, and the binomial distribution is a particularly useful example of type I. A little later in the chapter, in Table 6, the student will find an empirical frequency distribution together with the corresponding binomial distribution which adequately describes the data. Theoretical continuous distributions of type III will be described in Chapter 5.

We now return to a discussion of some of the properties of the binomial distribution.

4.5 The mean and variance of the binomial distribution

In Chapter 2 we saw how to calculate the sample mean of an empirical frequency distribution. If the values $x_1, x_2, \ldots, x_N$ occur with frequencies $f_1, f_2, \ldots, f_N$ then the sample mean is given by

$$\bar{x} = \frac{\sum\limits_{i=1}^{N} f_i x_i}{\sum\limits_{i=1}^{N} f_i}.$$

This quantity is an estimate of the *true population mean*, which is usually denoted by μ. In the case of the binomial distribution, the theoretical mean depends on the values of the binomial parameters, n and p. If a series of n trials is made, each of which has a probability p of being a success, then the average or mean number of successes is given by

$$\mu = np.$$

For example, if a penny is tossed 20 times, the average number of heads observed will be $10 = 20 \times \frac{1}{2}$, since $n = 20$ and $p = \frac{1}{2}$. Of course in a single experiment we may observe any integer between 0 and 20. The above result says that in the long run the average number of heads observed will be 10.

It is very important to understand the difference between the theoretical mean, μ, and the sample mean, $\bar{x}$. In most practical situations the theoretical mean is unknown and the sample mean is used to estimate it (see Chapter 6). But the two quantities will almost always be different.

To illustrate this point we will describe an experiment in which the theoretical mean is actually known. Suppose that each person in a group of people randomly selects ten telephone or car numbers. Then there is a probability $p = \frac{1}{2}$ that any number will be even. The actual total of even numbers that 1 person selects may be any integer between 0 and 10 and the probability of getting exactly r even numbers is given by the binomial distribution with $n = 10$ and $p = \frac{1}{2}$. These probabilities are given in Table 6. The experiment was carried out by 30 people and the number of people who chose r even numbers out of 10 is also recorded in Table 6. These observed frequencies can be compared with the binomial probabilities by multiplying the latter by the sample size, 30, to give a set of expected frequencies. Comparing the two distributions by eye, it can be seen that the binomial distribution gives a reasonably good fit to the observed frequency distribution.

The theoretical mean of this binomial distribution is given by $np = 5$. Thus a person will choose five even numbers on average.

Table 6
Theoretical and Observed Distributions of Even Telephone Numbers

r	P_r	Expected frequency $= P_r \times 30$	Observed frequency	Number of observed even numbers
0	0·001	0·0	0	0
1	0·010	0·3	0	0
2	0·044	1·3	2	4
3	0·117	3·5	3	9
4	0·205	6·2	5	20
5	0·246	7·4	9	45
6	0·205	6·2	3	18
7	0·117	3·5	6	42
8	0·044	1·3	1	8
9	0·010	0·3	1	9
10	0·001	0·0	0	0
		30·0	30	155

The sample mean of the observed frequency distribution can be obtained by dividing the total number of observed even numbers by thirty. This gives

$$\bar{x} = \tfrac{155}{30} = 5 \cdot 17.$$

Thus there is a difference of 0·17 between the sample mean and the theoretical mean. However the values are close enough together to demonstrate that the observed distribution is very similar to the theoretical binomial distribution. In fact in Chapter 6 we shall see that the sample mean, $\bar{x}$, tends to get closer to the theoretical mean as the sample size is increased. In other words if 1000 people each randomly select ten numbers then the average number of even numbers will probably be much closer to five.

4.5.1 *The expected value*

The mean of a probability distribution is not always as easy to find as that of the binomial distribution and so we will give a formal definition of the mean of a distribution based on the expected value symbol E. The idea of the expected value of a random variable is an extremely useful one and can be used to calculate many other quantities (see section 6.2).

If X is a discrete random variable which can take values $x_1, x_2, \ldots, x_N$ with probabilities $p_1, p_2, \ldots, p_N$ then the average or *expected* value of X is given by

$$E(X) = \sum_{i=1}^{N} x_i p_i. \qquad \textbf{4.1}$$

It may not be immediately obvious that this formula will give a quantity which is the average value of X in a long series of trials. However there is a clear analogy with the formula for the sample mean if we use the fact that $\sum_{i=1}^{N} p_i = 1$ and write the above formula as

$$E(X) = \frac{\displaystyle\sum_{i=1}^{N} x_i p_i}{\displaystyle\sum_{i=1}^{N} p_i}.$$

We can check this formula by calculating the expected value of a binomial random variable which we already know is np. From equation **4.1** we have

$$E(X) = \sum_{r=0}^{n} {}^{n}C_r p^r (1-p)^{n-r} r$$

$$= np \sum_{r=0}^{n-1} {}^{n-1}C_r p^r (1-p)^{n-r-1}$$

$$= np[p + (1-p)]^{n-1}$$

$$= np.$$

Example 8

Toss two coins and let X = number of heads. It is clear that the expected number of heads will be one. This is confirmed by applying the above formula which gives

$$E(X) = 0 \times P(X = 0) + 1 \times P(X = 1) + 2 \times P(X = 2)$$

$$= 1 \times \tfrac{1}{2} + 2 \times \tfrac{1}{4}$$

$$= 1.$$

Example 9

A gambler wins £20 with probability 0·1 and loses £1 with probability 0·9. What is his expected win?

Let X = amount won.

Then $X = +20$ with probability 0·1

$X = -1$ with probability 0·9.

Thus $E(X) = 20 \times 0·1 - 1 \times 0·9$

$= 1·1$.

Note that the expected value need not be a possible value of the random variable. Thus in Example 9 the gambler cannot win £1·1 in any particular game. The result says that in a long series of games he will win £1·1 per game on average.

4.5.2   *Variance*

The idea of an expected value is also extremely useful when calculating the theoretical variance, and hence the theoretical standard deviation, of a probability distribution.

We have already seen in Chapter 2 how to calculate the sample variance of a set of data. If n observations $x_1, \ldots, x_n$ are taken from a population which has mean μ, the sample variance is given by

$$s^2 = \frac{\sum\limits_{i=1}^{n} (x_i - \mu)^2}{n}.$$

This quantity is an estimate of the *true population variance*, which is usually denoted by σ^2. Let us suppose that X is a discrete random variable, with mean μ, which can take values $x_1, x_2, \ldots, x_N$ with probabilities $p_1, p_2, \ldots, p_N$. Then by analogy with the formula for s^2 we have

variance $(X) = \sigma^2$

$$= \sum_{i=1}^{N} (x_i - \mu)^2 p_i. \qquad \textbf{4.2}$$

The variance is a measure of the spread of the distribution around its expected value μ. In terms of the expected value symbol we have variance $(X) = E[(X - \mu)^2]$.

In other words the variance of X is the average or expected value of $(X - \mu)^2$ in a long series of trials.

From equation **4.2** the theoretical variance of the binomial distribution is given by

$$\sum_{r=0}^{n} (r - np)^2 \ {}^nC_r p^{\ r}(1-p)^{n-r},$$

which after a lot of algebra gives the simple formula $np(1 - p)$. From this result it follows that the standard deviation of the binomial distribution is given by $\sqrt{[np(1-p)]}$.

4.6 The Poisson distribution

Another important discrete distribution is the *Poisson* distribution, which is named after a French mathematician. The probability distribution of a Poisson random variable is given by

$$P(r) = \frac{e^{-\mu}\mu^r}{r!} \qquad (r = 0, 1, 2, \ldots \quad \mu > 0).$$

These probabilities are non-negative and sum to one.

$$\sum_{r=0}^{\infty} P(r) = \sum_{r=0}^{\infty} \frac{e^{-\mu}\mu^r}{r!}$$

$$= e^{-\mu} \sum_{r=0}^{\infty} \frac{\mu^r}{r!}$$

$$= e^{-\mu} e^{+\mu}$$

$$= 1.$$

Thus the probabilities form a discrete probability distribution. These probabilities are easy to calculate from $P(0) = e^{-\mu}$ using the recurrence relationship

$$P(r+1) = \frac{\mu P(r)}{(r+1)}.$$

The probabilities are also tabulated in Pearson and Hartley (1966) for $\mu = 0{\cdot}1(0{\cdot}1)15{\cdot}0$.

The Poisson distribution has two main applications; firstly for describing the number of 'accidents' which occur in a certain time interval, and secondly as a useful approximation to the binomial distribution when the binomial parameter p is small. We begin with the former situation.

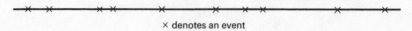

× denotes an event

Figure 15

Let us suppose that events occur randomly as shown in Figure 15 and that there are λ events on average in a unit time interval. This means that in a time interval of length t there will be λt events on average. But the actual number observed in one particular time interval may be any non-negative integer. A typical situation of this type was presented in Example 2, Chapter 1, where the arrival of a cosmic particle can be thought of as an 'accident' or event. The number of particles striking an apparatus in successive periods of one minute varied between nought and four. The reader must realize that we cannot predict exactly how many particles will arrive in a particular time interval. What we can do is to predict the pattern of arrivals in a large number of such time intervals. We will show that the number of accidents, in a time interval of length t, is a Poisson random variable with parameter $\mu = \lambda t$. The proof is rather sophisticated and the student may prefer to omit the following subsection at the first reading.

4.6.1 The Poisson model

In a very small time interval of length Δt, let us suppose that the probability of observing one accident is given by $\lambda \Delta t$ and the chance of observing more than one accident is negligible. In addition let us suppose that the numbers of accidents in two different time intervals are independent of one another.

Let $P(r, t)$ = probability of observing exactly r accidents in a time interval of length t. Then,

$$P(0, \Delta t) = 1 - \lambda \Delta t \quad \text{(from the first assumption)}$$

and

$$P(0, t + \Delta t) = P(0, t)P(0, \Delta t) \quad \text{(by the assumption of independence)}$$

$$= P(0, t)(1 - \lambda \Delta t).$$

Rearranging this equation we obtain

$$\frac{P(0, t+\Delta t)-P(0, t)}{\Delta t} = -\lambda P(0, t).$$

The left hand side of this equation can be recognized as the definition of

$$\frac{dP(0, t)}{dt} \quad \text{as} \quad \Delta t \to 0.$$

This simple first order differential equation for $P(0, t)$ can be solved to give

$$P(0, t) = e^{-\lambda t},$$

using the initial condition $P(0, 0) = 1$.

A similar differential equation for $P(1, t)$ can be obtained by noting that

$$P(1, t+\Delta t) = P(1, t)P(0, \Delta t)+P(0, t)P(1, \Delta t)$$

$$= P(1, t)(1 - \lambda \Delta t)+e^{-\lambda t}\lambda \Delta t.$$

This gives

$$\frac{P(1, t+\Delta t)-P(1, t)}{\Delta t} = \frac{dP(1, t)}{dt}$$

$$= -\lambda P(1, t)+\lambda e^{-\lambda t},$$

which can be solved to give

$$P(1, t) = e^{-\lambda t}\lambda t,$$

using the initial condition $P(1, 0) = 0$.

In a similar way, we can find $P(2, t)$, $P(3, t) \ldots$. Generally we find

$$P(r, t) = \frac{e^{-\lambda t}(\lambda t)^r}{r!} \qquad (r = 0, 1, 2, \ldots).$$

If we put $\mu = \lambda t$

$$= \text{(average number of accidents in time } t),$$

then we have a Poisson distribution as previously defined.

This model for the Poisson distribution depends on a random physical mechanism called a *Poisson process*, further details of which can be obtained from any book on *stochastic* (that is, random) processes (see the Further Reading list).

4.6.2 *Properties of the Poisson distribution*

(i) In the Poisson process, the average number of accidents in time t is given $\mu = \lambda t$. Thus μ must be the *mean* of the Poisson distribution. The mean can also be calculated using equation **4.1**.

$$\text{Mean} = \sum_{r=0}^{\infty} rP(r) = \sum_{r=1}^{\infty} \frac{e^{-\mu}\mu^r}{(r-1)!}$$

$$= \mu \sum_{r=0}^{\infty} \frac{e^{-\mu}\mu^r}{r!}$$

$$= \mu.$$

(ii) An important feature of the Poisson distribution is that the *variance* is equal to the mean, μ. Thus the standard deviation is given by $\sqrt{\mu}$. We find

$$\text{variance} = E[(X-\mu)^2]$$

$$= \sum_{r=0}^{\infty} (r-\mu)^2 P(r) \qquad \text{(from equation **4.2**)}$$

$$= \sum_{r=0}^{\infty} \{r(r-1)+r-2\mu r+\mu^2\}P(r)$$

$$= \mu^2+\mu-2\mu^2+\mu^2 \qquad \text{(after some algebra)}$$

$$= \mu.$$

(iii) The Poisson distribution is a useful approximation to the binomial distribution when the binomial parameter p is small. The binomial distribution has mean np and variance $np(1-p)$. Thus when p is small we find

$$\text{variance} \simeq np = \text{mean},$$

as for the Poisson distribution. Then it can be shown (see for example, Hoel, 1962) that if $n \to \infty$ and $p \to 0$ in such a way that the mean, $\mu = np$, stays fixed, then the binomial probabilities tend to the Poisson probabilities. This is of practical importance because it is much easier to calculate Poisson probabilities than the corresponding binomial probabilities. Generally speaking if n is large (>20) and the mean, $\mu = np$, is small (<5) then the approximation will be sufficiently accurate.

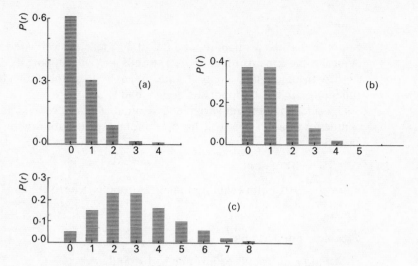

Figure 16   Poisson distributions
(a) $\mu = 0\cdot5$ (b) $\mu = 1$ (c) $\mu = 3$

(iv) Figure 16 shows three examples of Poisson distributions, with different values for μ, to show the types of distribution which can occur.

Example 10

Suppose that the number of dust particles per unit volume in a certain mine is randomly distributed with a Poisson distribution and that the average density is μ particles per litre.

A sampling apparatus collects a one-litre sample and counts the number of particles in it. If the true value of μ is six, what is the probability of getting a reading less than two?

Let $X =$ (number of particles in a one-litre sample).

$$P(X = r) = e^{-6}\frac{6^r}{r!} \qquad (r = 0, 1, 2, \ldots).$$

Therefore

$$P(X < 2) = P(X = 0) + P(X = 1)$$

$$= e^{-6} + 6\,e^{-6}$$

$$= 0\cdot0174.$$

Example 11

Suppose that the probability of a defect in a mile of steel wire is 0·01. A steel cable consists of a hundred strands and will support its design load with ninety-nine good strands. What is the probability that a mile-long cable will support its design load?

Consider the random variable X = number of defective strands in a mile-long cable. This will have a binomial distribution with parameters

$$n = 100 \text{ (strands)}$$

and $p = 0·01 = $ (probability that any strand is defective).

Then

P(cable is not defective)

$$= P(\text{no bad strands or just one bad strand})$$

$$= {}^{100}C_0(0·01)^0(0·99)^{100} + {}^{100}C_1(0·01)(0·99)^{99}$$

$$= 0·367 + 0·370$$

$$= 0·737.$$

As n is large and p is small it is possible to approximate this binomial distribution with a Poisson distribution which has the same mean

$$\mu = np = 1.$$

The probability that the cable has r defective strands is given approximately by $e^{-1}/r!$. Thus the probability that the cable is not defective is given approximately by

$$e^{-1} + e^{-1} = 0·368 + 0·368$$

$$= 0·736.$$

The answers obtained from the binomial and Poisson distributions are virtually identical, and since it is much easier to calculate e^{-1} than $0·99^{100}$, it is preferable to use the Poisson distribution approximation.

Example 12

The number of cosmic particles striking an apparatus in forty consecutive periods of one minute were given in Example 2, Chapter 1. On the assumption that cosmic particles arrive randomly at a constant over-all rate, one expects the Poisson distribution to describe the frequency distribution of the number of particles arriving in a one-minute period.

The total number of particles observed is

$$13 + (2 \times 8) + (3 \times 5) + (4 \times 1) = 48.$$

(The average number of particles observed per minute) $= \frac{48}{40} = 1\cdot2$.

The Poisson distribution with this mean is given by

$$P(r) = e^{-1\cdot2}\frac{(1\cdot2)^r}{r!} \qquad (r = 0, 1, 2, \ldots)$$

= (probability of observing r particles in a one-minute period if $\mu = 1\cdot2$).

These probabilities are tabulated in column 3 of Table 7.

Table 7

Number of particles (r) in a one-minute period	Number of periods with r particles	Poisson probabilities	Poisson frequencies
0	13	0·301	12·0
1	13	0·361	14·4
2	8	0·216	8·7
3	5	0·087	3·5
4	1	0·026	1·0
5+	0	0·011	0·4
Total	40	1·0	40·0

Multiplying these probabilities by forty (the sample size) gives the frequencies which can be expected if the Poisson model is appropriate. These frequencies are tabulated in column 4 of Table 7. Comparing the observed and theoretical frequencies by eye, we see that there appears to be good agreement between the two distributions. This indicates that the Poisson model really is appropriate in this situation.

4.7 Bivariate discrete distributions

So far we have considered the discrete probability distribution for a single random variable. It is a straightforward matter to extend the ideas to deal with the situation in which we are interested in two random variables, X and Y, at the same time.

The *joint probability* that X will take a particular value x, and that Y will take a particular value y is denoted by

$$P(x, y) = P(X = x, Y = y).$$

This function is such that

$$\sum_{x,y} P(x, y) = 1.$$

The probability of obtaining a particular value of one random variable without regard to the value of the other random variable is called the *marginal probability*. Thus

$$P_X(x) = \sum_y P(x, y),$$

$$P_Y(y) = \sum_x P(x, y)$$

are the marginal probability distributions of the two random variables.

The two random variables are said to be independent if

$$P(x, y) = P_X(x)P_Y(y) \qquad \text{for all values of } x \text{ and } y.$$

The idea of independence can be clarified as follows. Suppose that we know that a particular value of Y, say y, has been observed, and we want to know the probability that a particular value of X will occur. We write

$$P(x|y) = \text{probability that } X = x \text{ given that } Y = y.$$

We call this the conditional probability that X equals x given that the value of Y is y. From equation **3.4** we have

$$P(x|y) = \frac{P(x, y)}{P_Y(y)}.$$

Thus if X, Y are independent we have

$$P(x|y) = P_X(x)$$

and then the knowledge that a particular value of Y has been observed does not affect the probability of observing a particular value of X. The conditional distribution of X is then the same for all values of Y (and vice versa).

Example 13

Suppose the joint distribution of X and Y is as given below

		X 0	1	2	Marginal distribution of Y
	0	$\frac{1}{8}$	$\frac{1}{8}$	$\frac{1}{8}$	$\frac{3}{8}$
Y	1	$\frac{1}{4}$	0	$\frac{1}{8}$	$\frac{3}{8}$
	2	$\frac{1}{8}$	$\frac{1}{8}$	0	$\frac{1}{4}$
Marginal distribution of X		$\frac{1}{2}$	$\frac{1}{4}$	$\frac{1}{4}$	

The grand total of the joint probabilities is one. The row totals form the marginal distribution of Y; the column totals form the marginal distribution of X.

If $Y = 0$ the conditional distribution of X is given by

$$P(0|0) = \frac{1/8}{3/8} = 1/3,$$

$$P(1|0) = \frac{1/8}{3/8} = 1/3,$$

$$P(2|0) = \frac{1/8}{3/8} = 1/3.$$

These conditional probabilities also add up to one. The other conditional distributions of both X and Y can be found in a similar fashion. Since the above conditional distribution is not the same as the marginal distribution of X, the two random variables are not independent.

Exercises

1. Packets of food are filled automatically and the proportion of packets in a very large batch which are underweight is p. A sample size n is selected randomly from the batch and the probability that the sample contains exactly r defective packets ($r = 0, 1, 2, \ldots, n$) follows a certain probability distribution. Name this distribution and write down the probability that the sample contains exactly r defective packets.

For one particular process it has been found in the past that 2 per cent of the packets are underweight. An inspector takes a random sample of ten packets. Calculate

(a) the expected number of packets in the sample which are underweight,

(b) the probability that none of the packets in the sample are underweight.

(c) the probability that more than one of the packets in the sample is underweight.

2. Show that

(a) $^nC_{r-1} + {}^nC_r = {}^{n+1}C_r$,

(b) $1 + n + {}^nC_2 + \ldots {}^nC_n = 2^n$,

(c) $1 - n + {}^nC_2 - {}^nC_3 \ldots + (-1)^n {}^nC_n = 0$.

3. It has been found in the past that 4 per cent of the screws produced in a certain factory are defective. A sample of ten is drawn randomly from each hour's production and the number of defectives is noted. In what fraction of these hourly samples would there be at least two defectives? What doubts would you have if a particular sample contained six defectives?

4. One per cent of a certain type of car has a defective tail light. How many cars must be inspected in order to have a better than even chance of finding a defective tail light?

5. An electronic component is mass-produced and then tested unit by unit on an automatic testing machine which classifies the unit as 'good' or 'defective'. But there is a probability 0·1 that the machine will mis-classify the unit, so that each component is in fact tested five times and regarded as good if so classified three or more times. What now is the probability of a mis-classification?

6. A recent court decision supported a pilot's decision to continue to his destination with a four-engine jet aircraft after an engine failure at midrange on a two-hour flight, rather than land at the midway point. His argument was that the aircraft will fly on two engines and that the probability of two additional failures was quite small. If the one-hour reliability of a single engine is 0·9999, do you agree with the pilot's

decision? If the aircraft was a three-engine jet that will fly on one engine would you agree? Comment on the comparative reliability of the three- and four-engine jets in this situation.

7. The new 'Tigercat' sports car has an idle loping problem as about 10 per cent of the Tigercats have an unstable fluctuating engine speed when they are idling. An engineering 'fix' is put in a production pilot lot of a hundred cars.

(a) If the fix has no effect on the problem, how many cars would you expect to have the fault?

(b) If only two of the pilot lot have the idling fault, and the other ninety-eight cars are not defective, would you conclude that the fix has a significant effect? (Hint: Show that the probability of getting two or less defectives in a sample size of a hundred, given that the fix has no effect, is very small indeed.)

8. The average number of calls that a hospital receives for an ambulance during any half-hour period is 0·20. Considering a reasonable cost per ambulance and crew and presuming that any ambulance will return to the hospital in half an hour, how many ambulances would you recommend for this hospital? Comment on the idea of ambulance pools which are shared by several hospitals.

9. A manned interplanetary space vehicle has four engines each with reliability 0·99. Each engine has a failure detection system which may itself fail. If the engine does fail there is a conditional probability of 0·02 that a success will be signalled. If an engine fails and is not detected the result is catastrophic. However the mission can be completed with three engines if one engine fails and is detected. What is the probability of mission success? If there is no abort system or escape system, what action would you recommend if two failures are signalled? (Hint: Calculate probability of no engine failures plus one detected failure.)

10. It has been found in the past that one per cent of electronic components produced in a certain factory are defective. A sample of size one hundred is drawn from each day's production. What is the probability of getting no defective components by using

(a) the binomial distribution,

(b) the Poisson distribution approximation.

11. A construction company has a large fleet of bulldozers. The average number inoperative at a morning inspection due to breakdowns is two. Two standby bulldozers are available. If a bulldozer can always be mended within twenty-four hours of the morning inspection find the probability that at any one inspection

(a) no standby bulldozers will be required,

(b) the number of standby bulldozers will be insufficient.

12. There are many other discrete distributions apart from the binomial and Poisson distributions. For example the *hypergeometric* distribution arises as follows. A sample size n is drawn without replacement from a finite population size N which contains Np successes and $N(1-p)$ failures. Show that the probability of getting exactly x successes in the sample is given by

$$P(x) = \frac{^{Np}C_x \, ^{N(1-p)}C_{n-x}}{^{N}C_n} \qquad (x = 0, 1, 2, \ldots, n).$$

If N is very large compared with n, it can be shown that this distribution can be approximated by the binomial distribution with

$$P(x) \simeq {}^nC_x p^x (1-p)^{n-x} \qquad (x = 0, 1, 2, \ldots, n).$$

13. A series of items is made by a certain manufacturing process. The probability that any item is defective is a constant p, which does not depend on the quality of previous items. Show that the probability that the rth item is the first defective is given by

$$P(r) = p(1-p)^{r-1} \qquad (r = 1, 2, \ldots).$$

This probability distribution is called the *geometric* (or *Pascal*) distribution. Show that the sum of the probabilities is equal to one as required; also show that the mean of the distribution is equal to $1/p$.

References

HOEL, P. G. (1962), *Introduction to Mathematical Statistics*, Wiley, 3rd edn.
NATIONAL BUREAU OF STANDARDS (1950), *Tables of the Binomial Probability Distribution*, U.S. Government Printing Office, Washington, D.C.
PEARSON, E. S., and HARTLEY, H. O. (1966), *Biometrika Tables for Statisticians*, Cambridge University Press, 3rd edn.

Chapter 5
Continuous distributions

5.1 Definitions

In Chapter 4 we considered discrete distributions where the random variable can only take a discrete set of values. In this chapter we consider continuous distributions where the random variable can take any value in some specified interval.

In section 2.2 we described how observations on a continuous variate can be plotted as a histogram. As more and more observations are taken, and the class interval is made smaller, the histogram tends to a smooth curve called a frequency curve.

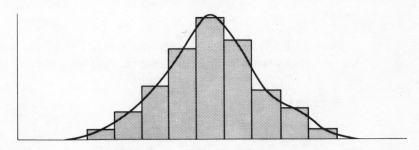

Figure 17   Frequency curve

If the height of the curve is standardized so that the area underneath it is equal to unity, then the graph is called a *probability curve*. The height of the probability curve at some point x is usually denoted by $f(x)$, and this function is called the *probability density function* (often abbreviated p.d.f.). This non-negative function is such that it satisfies the condition that the area under the probability curve is unity.

$$\int_{-\infty}^{\infty} f(x)\, dx = 1.$$

It is important to realize that $f(x)$ is *not* the probability of observing x. When the variate is continuous we can only find the probability of observing a value in a certain range. For example, $f(x_0)\Delta x$ is the probability of observing a value between x_0 and $x_0 + \Delta x$. In other words it is the area of the shaded strip in Figure 18.

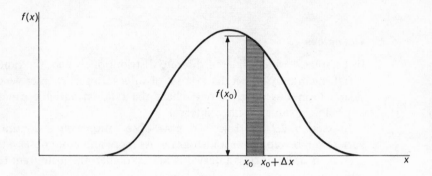

Figure 18

More generally the area between the verticals at x_1 and x_2 gives the probability that an observation will fall between x_1 and x_2.

$$\text{Probability } (x_1 < X < x_2) = \int_{x_1}^{x_2} f(x)\, dx.$$

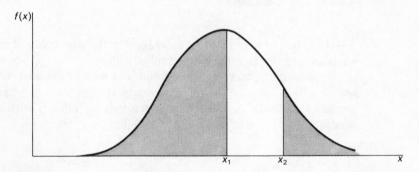

Figure 19

Example 1

A random variable is said to be *uniformly* distributed between *a* and *b* if it is equally likely to occur anywhere between *a* and *b*. Thus the height of the probability curve of the uniform distribution is a constant between *a* and *b*. Elsewhere the probability density function is zero.

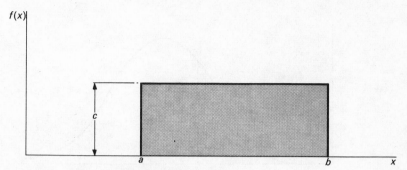

Figure 20   P.D.F. of uniform distribution

The total area under the curve is equal to $C(b-a)$. But this is equal to unity by definition. Thus we have

$$C = \frac{1}{(b-a)},$$

$$f(x) = \begin{cases} \dfrac{1}{(b-a)} & a < x < b, \\ 0 & \text{elsewhere.} \end{cases}$$

Another way of describing a probability distribution is to specify a function called the *cumulative distribution function* (often abbreviated c.d.f.). This function, usually denoted by $F(x)$, is defined by

$$F(x) = \text{probability}\,(X \leqslant x)$$

$$= \text{(probability of observing a value less than or equal to } x\text{)}.$$

This function is the theoretical counterpart of the cumulative frequency diagram which was described in section 2.2. From a mathematical point of view the cumulative distribution function is the best way of describing a distribution since it can be used for both discrete and continuous distributions. For this reason it is often simply called the distribution function.

For discrete distributions it is a step function which increases from zero to one. However, from a practical point of view, the function is much more useful for problems involving continuous variables, and the rest of the discussion will be concerned with continuous distributions.

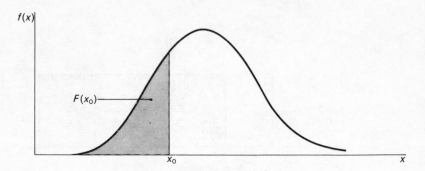

Figure 21

For such a distribution we find

$$F(x_0) = \int_{-\infty}^{x_0} f(x)\, dx$$

= (area under the probability curve to the left of x_0).

Thus $P(x_1 < X < x_2) = F(x_2) - F(x_1)$.

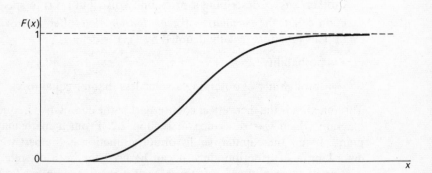

Figure 22   An S-shaped c.d.f.

The function must increase from zero to one since we have

$$F(-\infty) = 0$$

and

$$F(+\infty) = 1.$$

It is often S-shaped as in Figure 22.

Example 2

Find the cumulative distribution function of the uniform distribution (see Example 1). The random variable cannot take a value less than a.

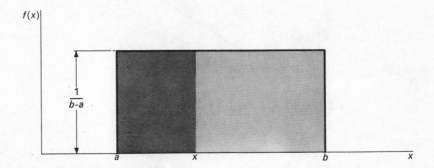

Figure 23

Thus we have

$$F(x) = 0 \qquad x < a.$$

In addition the random variable must always be less than or equal to b. Thus we have

$$F(x) = 1 \qquad x > b.$$

For values of x between a and b we have

$$F(x) = \frac{x-a}{b-a}.$$

Thus we have

$$F(x) = \begin{cases} 0 & x < a, \\[2mm] \dfrac{x-a}{b-a} & a \leqslant x \leqslant b, \\[2mm] 1 & x > b. \end{cases}$$

This function is illustrated below.

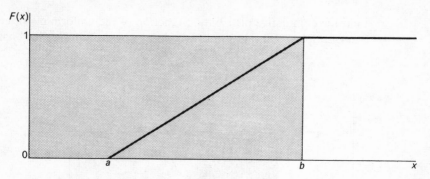

Figure 24   C.D.F. of uniform distribution

The c.d.f. of a continuous distribution describes it just as completely as the p.d.f. Thus the two functions are complementary, and they can be obtained from one another using the relations

$$F(x_0) = \int\limits_{-\infty}^{x_0} f(x)\,dx$$

and

$$f(x) = \frac{dF(x)}{dx}.$$

(Mathematical note: $F(x)$ may not be differentiable at isolated points and the p.d.f. is not defined at these points.)

5.2   The mean and variance of continuous distributions

By analogy with equation **4.1** the definition of the *mean* of a continuous distribution is given by

$$E(X) = \int_{-\infty}^{\infty} xf(x)\,dx. \qquad\qquad \textbf{5.1}$$

This value is the average or expected value of the random variable in a long series of trials. The summation sign in equation **4.1** is replaced with the integral sign.

By analogy with equation **4.2** the definition of the *variance* of a continuous distribution is given by

variance $(X) = E[(X - \mu)^2]$ where $\mu = E(X)$

$$= \int_{-\infty}^{\infty} (x - \mu)^2 f(x)\,dx. \qquad\qquad \textbf{5.2}$$

Example 3

Find the mean and variance of the uniform distribution which has p.d.f.

$$f(x) = \begin{cases} \frac{1}{2} & -1 < x < 1, \\ 0 & \text{elsewhere.} \end{cases}$$

By inspection the distribution is symmetric about the point $x = 0$, which must therefore be the mean of the distribution. This can be confirmed by equation **5.1** which gives

$$E(X) = \int_{-1}^{1} x\tfrac{1}{2}\,dx = 0.$$

Using equation **5.2** the variance is given by

$$\int_{-1}^{1} x^2 \tfrac{1}{2}\,dx = \tfrac{1}{3}.$$

5.3   **The normal distribution**

The *normal* or *Gaussian* distribution is the most important of all the distributions since it has a wide range of practical applications. It is sometimes called the bell-shaped distribution, a name which aptly describes the characteristic shape of many distributions which occur in practice. For example the histogram and frequency curve, which are plotted in Figure 6, are symmetric and shaped roughly like a bell.

The normal distribution is a mathematical model which adequately describes such distributions. The height of the normal probability curve is given by

$$f(x) = \frac{1}{\sqrt{(2\pi)}\sigma} e^{-(x-\mu)^2/2\sigma^2} \qquad -\infty < x < +\infty,$$

where μ, σ are parameters such that $-\infty < \mu < \infty$ and $\sigma > 0$.

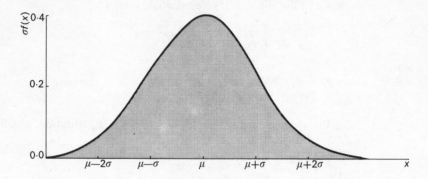

Figure 25   The normal distribution

The reader need not remember this rather complicated formula. A graph of the normal curve is given in Figure 25.

The p.d.f. is negligible for values of x which are more than 3σ away from μ.

It is worth pointing out that this p.d.f. has been chosen so that the total area under the normal curve is equal to one for all values of μ and σ. This can be shown by making the transformation $z = (x-\mu)/\sigma$ and using the standard integral

$$\int_{-\infty}^{\infty} e^{-z^2/2} \, dz = \sqrt{2\pi}$$

which is given in most tables of integrals.

The curve is symmetric about the point $x = \mu$, which must therefore be the mean of the distribution. It can also be shown, using equation **5.2**, that the variance of the normal distribution is equal to σ^2. Thus the standard deviation of the normal distribution is equal to σ.

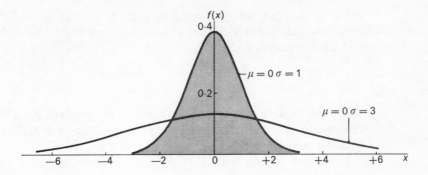

Figure 26   Two normal distributions

The shape of the normal curve depends on the standard deviation, σ, since the larger this is, the more spread out the distribution will be. Figure 26 shows two normal distributions, both of which have mean zero. However one has $\sigma = 1$ and the other has $\sigma = 3$.

Whatever the values of μ and σ, the normal distribution is such that about one observation in three will lie more than one standard deviation from the mean, and about one observation in twenty will lie more than two standard deviations from the mean. Less than one observation in 300 will lie more than three standard deviations from the mean.

In practical problems, the height of the normal curve is of little direct interest, and instead interest centres on the cumulative distribution function. For example, if the weights of a batch of screws are known to be approximately normally distributed with mean 2·10 grams and standard deviation 0·15 grams, we might want to find the proportion of screws which weigh less than 2·55 grams. This can be found from the cumulative distribution function.

$$F(x) = \text{probability } (X \leqslant x)$$

$$= \int_{-\infty}^{x} \frac{1}{\sqrt{(2\pi)}\sigma} e^{-(x-\mu)^2/2\sigma^2} \, dx$$

Unfortunately this integral cannot be evaluated as a simple function of x. Instead it must be integrated numerically, and the results can then be tabulated for different values of x. Table 1 in Appendix B tabulates the c.d.f. of a normal distribution with zero mean and a standard deviation of one. This particular normal distribution is often called the *standard normal distribution*. It turns out that the c.d.f. of any other normal distribution can also be obtained from this table by making the transformation

$$Z = \frac{X - \mu}{\sigma},$$

where the random variable X has mean μ and standard deviation σ. It is easy to show that the standardized variable Z has zero mean and a standard deviation of one, as required.

For a particular value, x, of X, the corresponding value of the standardized variable is given by

$$z = \frac{(x - \mu)}{\sigma},$$

and this is the number of standard deviations by which x departs from μ. The cumulative distribution function of X can now be found by using the relation

$$F(z) = \text{probability} \, (Z \leqslant z)$$

$$= \text{probability} \left(\frac{X - \mu}{\sigma} \leqslant z \right)$$

$$= \text{probability} \, (X \leqslant \mu + z\sigma).$$

The full table of $F(z)$ is given in Table 1, Appendix B. Some particularly useful values are also given in Table 8.

Table 8

z	0	1	2	2·5	3·0
$F(z)$	0·5	0·84	0·977	0·994	0·9987

Table 1, Appendix B, only tabulates $F(z)$ for positive values of z. Values of $F(z)$ when z is negative can be found by using the symmetry of the distribution. The two shaded areas in Figure 27 are equal in area.

$$\text{Probability} \, (Z \leqslant -z_0) = \text{probability} \, (Z > +z_0)$$

$$= 1 - \text{probability} \, (Z \leqslant +z_0),$$

Therefore $F(-z_0) = 1 - F(+z_0).$

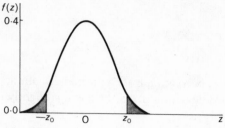

Figure 27

5.3.1 Notation

If a random variable, X, is normally distributed with mean μ and variance σ^2, we write

X is $N(\mu, \sigma^2)$.

The cumulative distribution function of the standard normal distribution, $N(0, 1)$, is often denoted by the special notation $\phi(z)$ to distinguish it from the c.d.f. of any other normal distribution. However in this book we will use the notation $F(z)$ on the understanding that the c.d.f. of a non-standard normal distribution will be described by probability$(X \leqslant x)$ and not $F(x)$.

Example 4

Find the probability that an observation from a normal distribution will be more than two standard deviations from the mean.

If X is $N(\mu, \sigma^2)$, then we want to find

probability $(X > \mu + 2\sigma) +$ probability $(X < \mu - 2\sigma)$.

By making the transformation $Z = (X - \mu)/\sigma$, it can be seen that this probability is the same as

probability $(Z > 2) +$ probability $(Z < -2)$.

But from Table 8 we have

$F(2) =$ probability $(Z \leqslant 2)$

$\quad\quad = 0.977$

therefore probability $(Z > 2) = 0.023$.

By symmetry (see Figure 27), this is equal to probability $(Z < -2)$.

Thus

$$\text{probability } (|Z| > 2) = 2 \times 0\cdot023$$

$$= 0\cdot046.$$

The probability that an observation from a normal distribution will be more than two standard deviations from the mean is slightly less than one in twenty.

Example 5

The individual weights of a batch of screws are normally distributed with mean $\mu = 2\cdot10$ grams and standard deviation $\sigma = 0\cdot15$ grams. What proportion of the screws weigh more than $2\cdot55$ grams?

In order to evaluate probability$(X > 2\cdot55)$ we make the transformation $Z = (X - \mu)/\sigma$. This gives

$$\text{probability } (X > 2\cdot55) = \text{probability } \left(\frac{X - 2\cdot10}{0\cdot15} > \frac{2\cdot55 - 2\cdot10}{0\cdot15} \right)$$

$$= \text{probability } (Z > 3)$$

$$= 1 - \text{probability } (Z \leqslant 3)$$

$$= 1 - F(3)$$

$$= 1 - 0\cdot9987 \qquad \text{(from Table 8)}$$

$$= 0\cdot0013.$$

Thus $0\cdot13$ per cent of the screws weigh more than $2\cdot55$ grams.

5.4 Uses of the normal distribution

(1) Many physical measurements are closely approximated by the normal distribution.

Generally speaking such measurements are of two types. Firstly, those in which the variation in the data is caused by observational error. If the error in measuring some unknown quantity is the sum of several small errors which may be positive or negative at random, then the normal distribution will usually apply. Secondly, measurements in which there is natural variation. For example, some biological measurements, such as the heights of different men, are approximately normally distributed.

Indeed non-normality is so rare that it is a useful clue when it does occur. Particular care is required when describing non-normal distributions particularly if the distribution is skewed.

(2) Some physical phenomena are not normally distributed but can be transformed to normality. For example the fatigue lives of a batch of electrical motors give a distribution which is skewed to the right. However if log(fatigue life) is plotted, then the distribution will be closer to a normal distribution.

(3) It can be shown that the normal distribution is a good approximation to the binomial distribution for large n, provided that p is not close to 0 or 1. (For example if $n > 20$ the approximation is valid for $0.3 < p < 0.7$. For larger values of n a wider range for p is permissible).

For a given binomial distribution, the corresponding normal distribution is found by putting

$$\mu = np \quad \text{and} \quad \sigma^2 = np(1-p).$$

This approximation is useful as the binomial distribution can be difficult to evaluate for large n (see Example 7). Thus to evaluate the binomial distribution

(a) if n is large and p is close to 0, use the Poisson approximation with $\mu = np$,

(b) if n is large and p is not close to 0 or 1, use the normal approximation as above,

(c) if n is small, then simply evaluate the binomial distribution.

(4) The normal approximation to the binomial distribution is a special case of the *central limit theorem*, which will be considered again in Chapter 6. Briefly this theorem says that if a series of samples, of size n, are taken from a population (not necessarily normal) with mean μ and standard deviation σ, then the sample means will form a distribution which tends to the normal distribution as n increases, whatever the population distribution. In Chapter 6 we will see that the distribution of sample means also has mean μ, but has a smaller standard deviation given by $\sigma/\sqrt{n}$.

(5) Any random variable formed by taking a linear combination of independent normally distributed random variables will itself be normally distributed.

93   Uses of the normal distribution

Example 6

The strengths of individual bars made by a certain manufacturing process are known to be approximately normally distributed with mean 24 and standard deviation 3. The consumer requires at least 95 per cent of the bars to be stronger than 20. Do the bars meet the consumer's specifications?

Let X = strength of a bar. In order to calculate probability $(X > 20)$, we standardize in the usual way to obtain

$$P(X > 20) = P\left[\frac{X-24}{3} > \frac{20-24}{3}\right]$$

$$= P\left[\frac{X-24}{3} > -1 \cdot 33\right].$$

The random variable $\frac{1}{3}(X-24)$ is approximately $N(0,1)$ so that this probability can be obtained from Table 1, Appendix B. We find

$$P(X > 20) = 0 \cdot 91.$$

Thus less than 95 per cent of the bars are stronger than 20 and so the bars do not meet the consumer's specifications.

Example 7

A die is tossed 120 times. Find the probability that a 'four' will turn up less than fifteen times.

Let X = number of 'fours' which turn up. This random variable will follow a binomial distribution with parameters $n = 120$ and $p = \frac{1}{6}$.

Thus $$P(X < 15) = \sum_{r=0}^{14} {}^{120}C_r\left(\frac{1}{6}\right)^r\left(\frac{5}{6}\right)^{120-r}$$

which is very tedious to sum.

However, as n is large and $p = \frac{1}{6}$ we can use the normal approximation with

$$\mu = np = 20$$

$$\sigma^2 = np(1-p) = 16 \cdot 67$$

$$\sigma = 4 \cdot 08.$$

Now for the discrete binomial distribution we have

$$P(X \leqslant 14) = P(X < 15),$$

which is not of course true for the continuous normal approximation. We compromise by finding $P(X < 14\frac{1}{2})$ with the normal approximation. (This is often called the *continuity correction*.)

$$P(X < 14\tfrac{1}{2}) = P\left(\frac{X-20}{4 \cdot 08} < \frac{14\frac{1}{2}-20}{4 \cdot 08}\right)$$

$$= P(Z < -1 \cdot 35)$$

$$= F(-1 \cdot 35)$$

$$= 1 - F(+1 \cdot 35)$$

$$= 0 \cdot 0885 \qquad \text{from Table 1, Appendix } \mathbf{B}.$$

5.5 Normal probability paper

The cumulative distribution function of the standard normal distribution is illustrated in Figure 28.

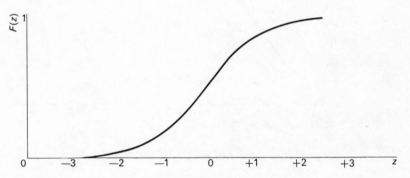

Figure 28   C.D.F. of standard normal distribution

It is possible to make a non-linear transformation of the vertical scale so that $F(z)$ will plot as a straight line.

Graph paper with scales such as those shown in Figure 29 is called *normal probability paper* and is used in the following way. Suppose we have a series of n observations which we suspect is normally distributed.

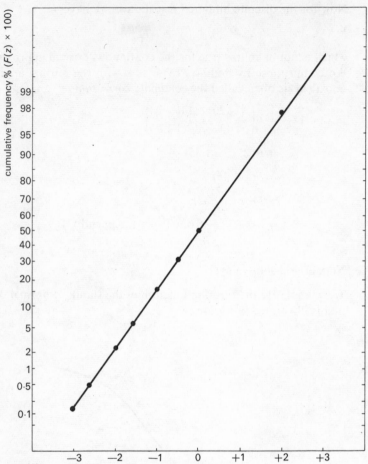

Figure 29   Normal probability paper

Arrange them in order of magnitude so that

$$x_1 \leqslant x_2 \leqslant \ldots \leqslant x_n.$$

At each point calculate the observed cumulative frequency

$$P(x_i) = \frac{\text{number of observations} \leqslant x_i}{n+1}$$

$$= \frac{i}{n+1}.$$

Note that there is a mathematical reason for putting $(n + 1)$ rather than n in the denominator. This also makes it possible to plot the point $P(x_n) = n/(n + 1)$ on the graph.

After choosing a suitable scale for the variate, x, the values of $P(x_i)$, $(i = 1$ to $n)$, can be plotted on the graph paper. If the data really is normal, the points will lie approximately on a straight line. Conversely if the data is not normal, as for example if the distribution is skewed, then the points will not lie on a straight line.

If the data does appear to be normal, the mean of the distribution can be estimated by fitting a straight line to the data and finding the value of x whose estimated cumulative frequency is 50 per cent. The standard deviation of the distribution can be estimated by finding the difference between the two values of x whose estimated cumulative frequencies are 50 per cent and 84 per cent respectively.

Example 8

The cumulative frequencies of the data from Example 1, Chapter 1, are given in Table 9.

Table 9

i	x_i	$P(x_i)$	i	x_i	$P(x_i)$	i	x_i	$P(x_i)$
1	989·4	0·032	11	998·6	0·355	21	1002·9	0·678
2	989·7	0·064	12	999·1	0·386	22	1003·1	0·710
3	992·8	0·097	13	999·2	0·418	23	1003·2	0·742
4	993·4	0·129	14	999·3	0·450	24	1004·5	0·774
5	994·7	0·161	15	1000·2	0·483	25	1006·5	0·805
6	995·3	0·193	16	1000·3	0·516	26	1006·7	0·838
7	996·4	0·225	17	1000·9	0·549	27	1007·6	0·870
8	996·5	0·257	18	1001·8	0·581	28	1008·7	0·902
9	997·9	0·290	19	1002·1	0·613	29	1012·3	0·934
10	998·1	0·322	20	1002·6	0·645	30	1014·5	0·977

These values are plotted on normal probability paper in Figure 30. The points lie roughly on a straight line and so the data is approximately normally distributed.

A straight line was fitted to the data by eye. From this line it can be seen that the value of x whose cumulative frequency is 50 per cent is 1000·6, and the difference between the two values of x whose cumulative

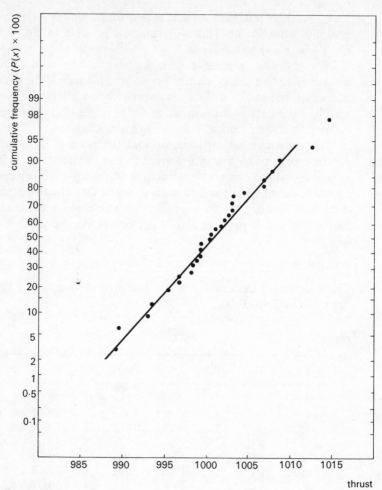

Figure 30

frequencies are 50 per cent and 84 per cent is 6·2. These values, which are estimates of the mean and standard deviation of the distribution, are very close to the sample mean and sample standard deviation as calculated in Chapter 2.

5.6 The exponential distribution

This is another useful continuous distribution. Its probability density function is given by

$$f(x) = \begin{cases} \lambda e^{-\lambda x} & x > 0 (\lambda > 0), \\ 0 & x < 0. \end{cases}$$

A typical example of an exponential distribution is given in Figure 31.

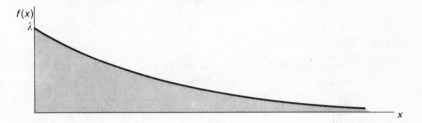

Figure 31   P.D.F. of exponential distribution

It is easy to show that the total area under the exponential curve is equal to unity. The probability of observing a value between a and b is given by

$$\int_a^b \lambda e^{-\lambda x} \, dx.$$

The cumulative distribution function of the exponential distribution is given by

$$F(x) = \begin{cases} 1 - e^{-\lambda x} & x \geqslant 0, \\ 0 & x < 0. \end{cases}$$

5.6.1   *Derivation from the Poisson process*

The exponential distribution can be obtained from the Poisson process (see section 4.6) by considering 'failure times' rather than the number of accidents (failures). If accidents happen at a constant rate λ per unit time, then the probability of observing no accidents in a given time t is given by $e^{-\lambda t}$.

Consider the random variable, T, which is the time that elapses after any time-instant until the next accident occurs. This will be greater than a particular value t provided that no accidents occur up to this point.

Thus   probability $(T > t)$ $= e^{-\lambda t}$,

therefore   probability $(T \leqslant t) = 1 - e^{-\lambda t}$.

This is the cumulative distribution function of the random variable T and we recognize that it is the c.d.f. of the exponential distribution. Thus times between accidents are distributed exponentially.

5.6.2   Mean and variance

If accidents happen at an average rate of λ per unit time then the average time between accidents is given by $1/\lambda$. This is the mean of the exponential distribution. It can also be obtained by calculating

$$E(X) = \int_0^\infty x\lambda\, e^{-\lambda x}\, dx = \frac{1}{\lambda}.$$

The variance of the exponential distribution is given by

$$\text{variance}(X) = E\left(X - \frac{1}{\lambda}\right)^2$$

$$= \int_0^\infty \left(x - \frac{1}{\lambda}\right)^2 \lambda\, e^{-\lambda x}\, dx$$

$$= \frac{1}{\lambda^2} \qquad \text{(after some algebra)}.$$

Example 9

The lifetime of a certain electronic component is known to be exponentially distributed with a mean life of 100 hours. What proportion of such components will fail before 50 hours?

As the mean life is 100 hours, we have $\lambda = 1/100$.

Thus the c.d.f. of the distribution of lifetimes is given by

$$F(t) = 1 - e^{-t/100}.$$

The probability that a component will fail before 50 hours is given by

$$F(50) = 1 - e^{-50/100}$$

$$= 0.393.$$

Thus the proportion of such components which fail before 50 hours is about 39 per cent.

5.6.3 Applications

The exponential distribution is sometimes used to describe the distribution of failure times when studying the reliability of a product. However the reader is warned that many products are such that the longer they survive the more likely they are to fail, in which case the exponential distribution will not apply. This subject is considered at length in Chapter 13.

The exponential distribution is also useful in the study of *queueing theory*, and a brief introduction to this subject is given here. Many situations can be described as a queueing process; this makes it possible to tackle problems in areas such as production engineering and road traffic flow. Two familiar examples of queues are customers making purchases at a shop, and calls arriving at a telephone exchange which only has a limited number of lines. A typical industrial situation occurs when one or more mechanics are available to repair machine breakdowns. The mechanic can be thought of as the server and a disabled machine as a customer. At a given time there may be several machines being repaired or waiting to be repaired, and these machines form the queue.

The simplest type of queue is a single-server queue, in which 'customers' arrive randomly at a single service station and receive attention as soon as all previous arrivals have been serviced. As customers arrive randomly, the number of arrivals in a given time-interval is a Poisson random variable. If the average arrival rate is denoted by λ and Δt is a very small time interval, we have

probability(new arrival between t and $t + \Delta t$) $= \lambda \Delta t$

and this is independent of previous arrivals.

The service time is occasionally constant, but more commonly it is a random variable and several distributions have been proposed to describe the distribution of service times. It is often found that the service times are exponentially distributed (see, for example, the discussion in Buffa, 1963). Denoting the exponential parameter by μ, the average service time is equal to $1/\mu$. The number of service completions in a given time interval is then a Poisson random variable, and so the probability that a service terminates between t and $t + \Delta t$ is given by $\mu \Delta t$ which is independent of previous service completions.

Let $P_r(t)$ denote the probability that there are r customers in the queue at time t (including the one being serviced). These probabilities will depend on the initial length of the queue at time $t = 0$. In order to calculate them it is necessary to solve a series of differential equations which are obtained as follows. The queue is empty at time $t + \Delta t$, either if it was empty at time t and there were no arrivals in Δt, or if the queue contained one customer at time t and there was a service completion in Δt. Using the fact that events in different time periods are independent and ignoring terms involving $(\Delta t)^2$ we have

$P_0(t + \Delta t) = P_0(t) \times$ probability (no arrivals in Δt)

$\qquad + P_1(t) \times$ probability (service completion in Δt)

$\qquad = P_0(t)(1 - \lambda \Delta t) + P_1(t)\mu \Delta t.$

This equation can be rearranged to give

$$\frac{P_0(t + \Delta t) - P_0(t)}{\Delta t} = -\lambda P_0(t) + \mu P_1(t),$$

where the left hand side can be recognized as the definition of

$$\frac{dP_0(t)}{dt} \quad \text{as} \quad \Delta t \to 0.$$

By a similar argument we have

$$\frac{dP_r(t)}{dt} = \lambda P_{r-1}(t) - (\lambda + u)P_r(t) + \mu P_{r+1}(t) \qquad (r = 1, 2, \ldots).$$

The general solution of this set of differential equations is given, for example, in Takacs (1962). An important special case is to find how the solutions behave as $t \to \infty$; in other words to find what is called the limiting distribution of the queue length. In fact, provided that $\lambda < \mu$, it is easy to show that this limiting distribution does exist and does not depend on the initial queue length. Then $[dP_r(t)]/(dt) \to 0$ as $t \to \infty$ and we find

$$P_r = \underset{t \to \infty}{\text{limit}} \, P_r(t)$$

$$= \left(\frac{\lambda}{\mu}\right)^r \left(1 - \frac{\lambda}{\mu}\right) \qquad (r = 0, 1, 2, \ldots).$$

This is an example of a geometric distribution (see exercise 13,

Chapter 4). The ratio λ/μ, which must be less than one to give a stable solution, is often called the *traffic intensity*. If $\lambda/\mu > 1$, the queue will tend to get longer and longer.

The above result enables us to calculate such useful characteristics of the system as the average queue size, the probability that the server is busy and the average waiting time of a customer. For example, the average queue size is given by

$$\sum_{r=0}^{\infty} rP_r = \frac{\lambda}{(\mu - \lambda)}$$

and the probability that the server is busy is given by

$$1 - P_0 = \frac{\lambda}{\mu}.$$

Some references on queueing theory are given in the bibliography.

5.7 Bivariate continuous distributions

As in section 4.7, it is a straightforward matter to extend the ideas of a single continuous random variable to the situation in which we are interested in two random variables, X and Y, at the same time.

The joint probability density function, $f(x, y)$, is given by the following relationship.

probability $(x < X < x + \Delta x, y < Y < y + \Delta y) = f(x, y)\Delta x \Delta y.$

This function is such that

$$\int_{-\infty}^{\infty} \int_{-\infty}^{\infty} f(x, y) \, dx \, dy = 1.$$

The probability density function of one random variable without regard to the value of the other random variable is called the marginal probability density function. Thus

$$f_X(x) = \int_{-\infty}^{\infty} f(x, y) \, dy,$$

$$f_Y(y) = \int_{-\infty}^{\infty} f(x, y) \, dx$$

are the marginal probability density functions of the two random variables.

The two random variables are said to be independent if

$$f(x, y) = f_X(x) f_Y(y) \qquad \text{for all } x, y.$$

Exercises

1. Find the area under the standard normal curve ($\mu = 0, \sigma = 1$)
(a) outside the interval $(-1, +1)$,
(b) between -0.5 and $+0.5$,
(c) to the right of 1.8.

2. If X is $N(3, 2)$ find
(a) $P(X < 3)$,
(b) $P(X \leqslant 5)$,
(c) $P(X \leqslant 1)$.

3. The mean weight of 500 students at a certain college is 150 lb and the standard deviation 15 lb. Assuming that the weights are approximately normally distributed, estimate the proportion of students who weigh
(a) between 120 and 180 lb,
(b) more than 180 lb.

4. The lengths of a batch of steel rods are approximately normally distributed with mean 3.1 ft and standard deviation 0.15 ft. Estimate the proportion of rods which are longer than 3.42 ft.

5. If X is normally distributed with $\mu = 10$ and $\sigma = 2$, find numbers x_0, x_1 such that
(a) $P(X > x_0) = 0.05$,
(b) $P(X > x_1) = 0.01$.
Also find k such that

$$P(|X - \mu| > k) = 0.05.$$

6. Suppose that the lifetimes of a batch of radio components are known to be approximately normally distributed with mean 500 hours and standard deviation 50 hours. A purchaser requires at least 95 per cent of them to have a lifetime greater than 400 hours. Will the batch meet the purchaser's specifications?

7. An examination paper consists of twenty questions in each of which the candidate is required to tick as correct one of the three possible answers. Assume that a candidate's knowledge about any question may be represented as either (i) complete ignorance in which case he ticks at random or, (ii) complete knowledge in which case he ticks the correct answer.

How many questions should a candidate be required to answer correctly if not more than 1 per cent of candidates who do not know the answer to any question are to be allowed to pass? (Use the normal approximation to the binomial distribution.)

8. The lengths of bolts produced in a certain factory may be taken to be normally distributed. The bolts are checked on two 'go–no go' gauges so that those shorter than 2·983 in. or longer than 3·021 in. are rejected as 'too-short' or 'too-long' respectively.

A random sample of $N = 300$ bolts is checked. If they have mean length 3·007 in. and standard deviation 0·011 in., what values would you expect for n_1, the number of 'too-short' bolts and n_2, the number of 'too-long' bolts?

A sample of $N = 600$ bolts from another factory is also checked. If for this sample we find $n_1 = 20$ and $n_2 = 15$, find estimates for the mean and standard deviation of the length of these bolts.

References

BUFFA, E. S. (1963), *Models for Production and Operations Management*, Wiley.
TAKACS, L. (1962), *Introduction to the Theory of Queues*, Oxford University Press.

Chapter 6
Estimation

In Chapters 4 and 5 a number of discrete and continuous distributions were studied. The importance of these theoretical distributions is that many physical situations can be described, at least approximately, by a mathematical model based on one of these distributions. A difficulty many students have at first is that of recognizing the appropriate model for a particular situation. This can only come with practice and the student is recommended to try as many exercises as possible.

We will now turn our attention to the major problem of statistical inference which is concerned with getting information from a sample of data about the population from which the sample is drawn, and in setting up a mathematical model to describe this population.

The first step in this procedure is to specify the *type* of mathematical model to be employed. This step is vital since any deductions will depend upon the validity of the model. The choice of the model depends upon a number of considerations, including any relevant theoretical facts, any prior knowledge the experimenter has, and also a preliminary examination of the data. The model often assumes that the population of possible observations can be described by a particular theoretical distribution and we are now in a position to consider such models.

Statistical inference can be divided into two closely related types of problems; the *estimation* of the unknown parameters of the mathematical model and the *testing of hypotheses* about the mathematical model. The second of these problems will be considered in Chapter 7.

6.1 Point and interval estimates

It is most important to distinguish between the true population parameters and the sample estimates. For example, suppose that a sample of n observations $x_1, x_2, \ldots, x_n$, has a symmetric distribution similar to the distribution of the data in Example 1, Chapter 1. Then a suitable mathematical model is that each observation is randomly selected from a normal distribution of mean μ and standard deviation σ. Then the sample mean $\bar{x}$ is an intuitive estimate of the population mean μ and the sample standard deviation,

$$s = \sqrt{\frac{\sum\limits_{i=1}^{n} (x_i - \bar{x})^2}{n-1}}$$

is an intuitive estimate of the population standard deviation σ.

There are two types of estimators in common use. An estimate of a population parameter expressed by a single number is called a *point estimate*. Thus in the above example $\bar{x}$ is a point estimate of μ.

However, a point estimate gives no idea of the precision of the estimate. For example, it does not tell us the largest discrepancy between $\bar{x}$ and μ which is likely to occur. Thus it is often preferable to give an estimate expressed by two numbers between which the population parameter is confidently expected to lie. This is called an *interval estimate*.

Before discussing methods of finding point and interval estimates, we must decide what we mean by a 'good' estimate and for this we need to understand the idea of a *sampling distribution*.

Given a set of data, $x_1, x_2, \ldots, x_n$, a variety of statistics can be calculated such as $\bar{x}$ and s^2. If we now take another sample of similar size from the *same* population then slightly different values of $\bar{x}$ and s^2 will result. In fact if repeated samples are taken from the same population then it is convenient to regard the statistic of interest as a random variable and its distribution is called a sampling distribution. Strictly speaking the distribution of any random variable is a sampling distribution, but the term is usually reserved for the distribution of statistics like $\bar{x}$ and s^2.

6.2 Properties of the expected value

In order to find the sampling distribution of a statistic like $\bar{x}$, we shall have to extend our knowledge of the expected value of a random variable. This concept was introduced in Chapter 4 in order to calculate the theoretical mean and variance of a given distribution, but it is also useful for carrying out many other operations.

We begin by repeating the formal definition of the expected value of a simple random variable. If the discrete random variable X can take values $x_1, x_2, \ldots, x_N$, with probabilities $p_1, p_2, \ldots, p_N$, then the mean or expected value of the random variable is defined as

$$E(X) = \sum_{i=1}^{N} x_i p_i \quad \text{(see section 4.5).}$$

The corresponding definition for a continuous random variable is given by

$$E(X) = \int_{-\infty}^{\infty} x f(x)\, dx \quad \text{(see section 5.2).}$$

More generally we are interested in the expected value of random variables related to X. For example, we might want to know the expected value of $2X$ or X^2. The variance of X is of particular importance. This is given by

$$\text{variance}(X) = E[(X - \mu)^2] \quad \text{where } \mu = E(X).$$

These expected values can be found from the following general definition. If $g(X)$ is a function of the random variable, X, then the *expected value* or *expectation* of $g(X)$ is given by

$$E[g(X)] = \begin{cases} \displaystyle\sum_{i=1}^{N} g(x_i) p_i & \text{for the discrete case,} \\[2ex] \displaystyle\int_{-\infty}^{\infty} g(x) f(x)\, dx & \text{for the continuous case.} \end{cases}$$

The idea of expectation can also be extended to apply to problems involving more than one random variable. For example, we might want to calculate the expected value of $X + Y$, where X and Y are two random variables. If $g(X, Y)$ is a function of the random variables, X and Y, then the expected value or expectation of $g(X, Y)$ is given by

$$E[g(X, Y)] = \begin{cases} \displaystyle\sum_{x,y} g(x, y) P(x, y) & \text{for the discrete case,} \\[2ex] \displaystyle\int_{x,y} g(x, y) f(x, y)\, dx\, dy & \text{for the continuous case.} \end{cases}$$

In the discrete case $P(x, y)$ is the joint probability that X is equal to x and Y is equal to y, and in the continuous case $f(x, y)$ is the joint probability density function. The summation (or integral) is taken over all possible combinations of x and y.

In proving the following results we only consider the discrete case. The proof in the continuous case follows immediately. The quantities b and c are constants.

6.2.1   If X is any random variable then

$$E(bX + c) = bE(X) + c.$$ **6.1**

Proof

$$E(bX + c) = \sum_{i=1}^{N} (bx_i + c)p_i$$

$$= b \sum_{i=1}^{N} x_i p_i + c \sum_{i=1}^{N} p_i$$

$$= bE(X) + c.$$

6.2.2   If X and Y are any two random variables then

$$E(X + Y) = E(X) + E(Y).$$ **6.2**

Proof

$$E(X + Y) = \sum_{x,y} (x + y)P(x, y)$$

$$= \sum_{x,y} xP(x, y) + \sum_{x,y} yP(x, y)$$

$$= \sum_{x} x \left[\sum_{y} P(x, y) \right] + \sum_{y} y \left[\sum_{x} P(x, y) \right]$$

$$= \sum_{x} xP_X(x) + \sum_{y} yP_Y(y),$$

where $P_X(x)$, $P_Y(y)$ are the marginal probabilities of x and y respectively (see section 4.7).

This completes the proof as

$$\sum xP_X(x) = E(X) \quad \text{and} \quad \sum yP_Y(y) = E(Y).$$

This result can be extended to any number of random variables. We find

$$E(X_1 + X_2 + \ldots + X_k) = E(X_1) + \ldots + E(X_k).$$

6.2.3　If X is any random variable then

variance $(cX) = c^2$ variance (X).　　　　　**6.3**

Proof. Let $E(X) = \mu$, so that $E(cX) = c\mu$ by **6.1**. Then

$$\text{variance } (cX) = E[(cX - c\mu)^2]$$
$$= E[c^2(X - \mu)^2]$$
$$= c^2 E[(X - \mu)^2]$$
$$= c^2 \text{ variance } (X).$$

6.2.4　If X and Y are *independent* random variables, then

$$E(XY) = E(X)E(Y).$$　　　　　**6.4**

Proof

$$E(XY) = \sum_{x,y} xyP(x, y).$$

But X, Y are independent, so $P(x, y)$ can be factorized to give

$$P(x, y) = P_X(x)P_Y(y)$$
$$E(XY) = \sum_{x,y} xyP_X(x)P_Y(y)$$
$$= \left(\sum_x xP_X(x)\right)\left(\sum_y yP_Y(y)\right)$$
$$= E(X)E(Y).$$

6.2.5　If X and Y are *independent* random variables, then

variance $(X + Y) =$ variance $(X) +$ variance (Y).　　　　　**6.5**

This result is one of the many reasons why statisticians prefer the variance (or its square root the standard deviation) as a measure of the spread of a distribution in most situations.

Proof. Let　$E(X) = \mu_1$　and　$E(Y) = \mu_2$.

$$E(X + Y) = \mu_1 + \mu_2 \qquad \text{using } \textbf{6.2}.$$

$$\text{Variance } (X + Y) = E[(X + Y - \mu_1 - \mu_2)^2]$$
$$= E[(X - \mu_1)^2] + E[(Y - \mu_2)^2] + 2E[(X - \mu_1)(Y - \mu_2)].$$

But the third term in this expression is given by

$$E[XY - \mu_1 Y - \mu_2 X + \mu_1 \mu_2]$$

$$= E(X)E(Y) - \mu_1 E(Y) - \mu_2 E(X) + \mu_1 \mu_2 \qquad \text{using } \mathbf{6.4}$$

$$= 0.$$

Thus variance $(X + Y)$ = variance (X) + variance (Y).

This result can be extended to any number of *independent* random variables. We find

$$\text{variance } (X_1 + X_2 + \ldots + X_k) = \text{variance } (X_1) + \ldots + \text{variance } (X_k).$$

Note particularly that 6.2.4 and 6.2.5 only apply to independent random variables but that the other results apply to any random variables.

The following results are also useful. The proofs are left to the reader.

Variance $(X) = E(X^2) - [E(X)]^2.$ **6.6**

Variance $(X + c)$ = variance (X). **6.7**

6.3 The sampling distribution of $\bar{x}$

The most important sampling distribution is that of the sample mean. If repeated samples of the same size are taken from the same population, then the sample means will vary somewhat from sample to sample, and the values will form the sampling distribution of $\bar{x}$ (so called because it is a distribution obtained by repeated sampling).

As an example let us look once again at the data of Example 1, Chapter 1, in which $\bar{x} = 1000.6$ and $s = 6.0$ for thirty observations. If we split the observations into six groups of five observations and calculate the mean of each group, we find that the group means are 998.3, 1001.4, 1001.8, 998.4, 1004.5, and 999.2. These six values are the means of samples, size five. The mean of the group means is again 1000.6, but the standard deviation of the group means is only 2.4. Thus the group means are much closer together than the original observations. This confirms the intuitive idea that the more observations are taken the more accurate the mean of these observations will be. The mean of all thirty observations will be even more accurate than the group means.

We now state the following important theorem.

Theorem 1

If random samples, size n, are taken from a distribution with mean μ and standard deviation σ, then the sample means will form a distribution with the same mean μ but with a smaller standard deviation given by $\sigma/\sqrt{n}$.

The standard deviation of this sampling distribution is usually called the *standard error of the sample mean* to distinguish it from the standard deviation of the parent distribution. In fact the standard deviation of any sampling distribution is usually called a standard error.

Proof

(a) *The mean.* We want to find the average value of the sample mean, $\bar{x}$, in a long series of trials. It is convenient to use the concept of the expected value. By definition we know that the expected value of any observation is equal to the population mean μ.

$$\text{Thus} \quad E(\bar{x}) = E\left\{\frac{(x_1 + x_2 + \ldots + x_n)}{n}\right\}$$

$$= E\left(\frac{x_1}{n}\right) + E\left(\frac{x_2}{n}\right) + \ldots + E\left(\frac{x_n}{n}\right) \qquad \text{using } \mathbf{6.2}$$

$$= \frac{\mu}{n} + \frac{\mu}{n} + \ldots + \frac{\mu}{n} \qquad \text{using } \mathbf{6.1}$$

$$= \mu.$$

Thus the mean value of the sampling distribution of $\bar{x}$ is equal to μ. Note that when we write $E(\bar{x})$ or $E(x_1)$, the quantity in the brackets is a random variable and not a particular sample value.

(b) *The standard error.*

$$\text{Variance } (\bar{x}) = \text{variance}\left(\frac{\sum x_i}{n}\right)$$

$$= \frac{1}{n^2}\text{variance } \left(\sum x_i\right) \qquad \text{using } \mathbf{6.3}$$

$$= \frac{1}{n^2}[\text{variance } (x_1) + \ldots + \text{variance } (x_n)] \qquad \text{using } \mathbf{6.5}$$

since successive observations are independent.
But   variance $(x_i) = \sigma^2$ for all i.

Thus variance $(\bar{x}) = \dfrac{1}{n^2}[\sigma^2 + \ldots + \sigma^2]$

$$= \frac{\sigma^2}{n}.$$

Thus the standard deviation or standard error of $\bar{x}$ is equal to $\sigma/\sqrt{n}$. This completes the proof.

Example 1

The percentage of copper in a certain chemical is to be estimated by taking a series of measurements on small random quantities of the chemical and using the sample mean percentage to estimate the true percentage. From previous experience individual measurements of this type are known to have no systematic error and to have a standard deviation of 2 per cent. How many measurements must be made so that the standard error of the estimated percentage is less than 0·6 per cent?

Assume that n measurements are made. The standard error of the sample mean will then be $2/\sqrt{n}$ per cent. If the required precision is achieved we must have

$$2/\sqrt{n} < 0.6$$

giving $\qquad n > 11{\cdot}1.$

As n must be an integer, at least twelve measurements must be made to achieve the required precision.

The student must remember Theorem 1, as it is one of the most important results in statistics. It is worth emphasizing that this theorem is independent of the parent distribution and holds whether it is normal, Poisson or any other. The question which now arises is what type of distribution will the sample mean follow? In Theorem 2 we consider the situation in which the parent distribution is normal.

Theorem 2

If random samples, size n, are taken from a normal distribution, the sampling distribution of $\bar{x}$ will also be normal, with the same mean and a standard error given by $\sigma/\sqrt{n}$.

What happens if the parent population is not normal? The rather surprising result is that provided reasonably large samples are taken (for example, $n > 30$), the sampling distribution of $\bar{x}$ will be approximately normal whatever the distribution of the parent population. This remarkable result known as the central limit theorem has already been mentioned in section 5.4. ·

Central limit theorem

If random samples, size n, are taken from a distribution with mean μ and standard deviation σ, the sampling distribution of $\bar{x}$ will be approximately normal with mean μ and standard deviation $\sigma/\sqrt{n}$, the approximation improving as n increases.

The proofs of both Theorem 2 and the central limit theorem are beyond the scope of this book, but can be found, for example, in Hoel (1964).

Example 2

The diameters of shafts made by a certain manufacturing process are known to be normally distributed with mean 2·500 cm and standard deviation 0·009 cm. What is the distribution of the sample mean of nine such diameters selected at random? Calculate the proportion of such sample means which can be expected to exceed 2·505 cm.

From Theorem 2 the sampling distribution of $\bar{x}$ will also be normal with the same mean 2·500 cm but with a standard deviation (or standard error) equal to $0·009/\sqrt{9} = 0·003$ cm.

In order to calculate probability ($\bar{x} > 2·505$) we standardize in the usual way to obtain

$$\text{probability}\left[\frac{\bar{x}-2·500}{0·003} > \frac{2·505-2·500}{0·003}\right]$$

$$= \text{probability}\left[\frac{\bar{x}-2·500}{0·003} > 1·66\right].$$

The random variable $(\bar{x}-2·500)/0·003$ will be $N(0, 1)$ so that the required probability can be obtained from Table 1, Appendix B. The proportion of sample means which can be expected to exceed 2·505 cm is 0·048.

Example 3

The sampling distribution of $\bar{x}$ is illustrated by constructing a distribution from successive samples of ten random numbers. The mean of each sample is calculated and the frequency distribution of the sample means is obtained.

In this example, a random number is such that it is equally likely to be any integer between zero and nine. Thus the population probability distribution is given by

$$P_r = \tfrac{1}{10} \qquad (r = 0, 1, \ldots, 8, 9).$$

This distribution is sometimes called the discrete uniform distribution.

Random numbers of this type have been tabulated in a number of books; see, for example, Kendall and Babington-Smith (1951), Fisher and Yates (1963) and the Rand Corporation (1955). Alternatively they can be generated with the simple device shown in Figure 32.

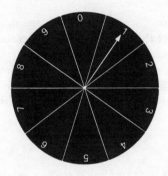

Figure 32   Random number generator

The circumference of the circle is divided into ten equal sections, numbered 0 to 9, so that when the arrow is spun it is equally likely to come to rest pointing at any one of the sections.

The arrow was spun ten times and the resulting values were 0, 8, 3, 7, 2, 1, 5, 6, 9, 9. The sample mean of these numbers is 5·0. The arrow was then spun ten more times and the resulting values were 2, 8, 7, 8, 1, 4, 5, 0, 3, 1, giving a sample mean equal to 3·9. This process was repeated forty times, giving forty sample means which are given below.

5·0	3·9	5·2	4·6	4·1	3·1	4·8	4·9	4·5	4·2
5·1	3·3	4·2	4·3	5·0	4·0	4·5	3·5	5·4	4·7
5·3	4·3	3·3	5·6	4·1	4·9	4·4	3·9	4·6	5·8
7·1	4·8	5·1	2·8	4·3	6·0	5·0	4·8	5·3	4·5

It is easier to inspect this data if we form the grouped sampling distribution as in the table below.

Grouped sampling distribution

Sample mean	Frequency
0·0–1·0	0
1·1–2·0	0
2·1–3·0	1
3·1–4·0	7
4·1–5·0	23
5·1–6·0	8
6·1–7·0	0
7·1–8·0	1
8·1–9·0	0

It is clear that this distribution has a smaller standard deviation than the parent distribution. In fact thirty-eight of the observations lie between 3·1 and 6·0. The mean of the parent population is 4·5. By inspection the mean value of the sampling distribution of $\bar{x}$ is also close to 4·5. Moreover the sampling distribution of $\bar{x}$ is much closer to a normal distribution than might have been expected with such relatively small samples.

6.4 The sampling distribution of s^2

Thus far we have concentrated on the sampling distribution of $\bar{x}$, but any other sample statistic will also have a sampling distribution. For example, if repeated random samples, size n, are taken from a normal distribution with variance σ^2, the statistic s^2 will vary from sample to sample and it can be shown that the statistic

$$\chi^2 = \frac{(n-1)s^2}{\sigma^2} = \frac{(x_1 - \bar{x})^2 + \ldots + (x_n - \bar{x})^2}{\sigma^2}$$

follows a distribution called the *chi-squared* or χ^2 distribution. The Greek letter χ is pronounced 'kigh' and spelt 'chi'. χ^2 is used rather than just χ to emphasize that the statistic cannot be negative. This distribution is related to the normal distribution (see Appendix) and depends upon a parameter called the 'number of degrees of freedom'. The term *degrees of freedom* (abbreviated d.f.) occurs repeatedly in statistics. Here it is simply a parameter which defines the particular χ^2 distribution. In the above situation, the distribution will have $(n-1)$d.f. This is the same as the denominator in the formula

$$s^2 = \frac{\sum(x_i - \bar{x})^2}{n-1}.$$

and we say that the estimate s^2 of σ^2 has $(n-1)$d.f. We will say more about this in section 6.5.

The p.d.f. of the χ^2 distribution is too complicated to give here, but the percentage points of the distribution are tabulated in Appendix B, Table 3, for different values of ν. The distribution is always skewed to the right and has a mean value equal to the number of degrees of freedom. For large values of ν, the distribution tends towards the normal distribution.

An example of a χ^2 distribution is shown in Figure 33. The percentage point $\chi^2_{\alpha,\nu}$ is chosen so that the proportion of the distribution, with ν d.f., which lies above it, is equal to α.

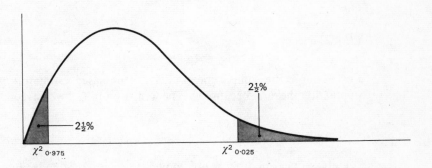

Figure 33   The χ^2 distribution

As $E(X^2_\nu) = \nu$

we have $E\left[\frac{(n-1)s^2}{\sigma^2}\right] = n-1,$

so that $E(s^2) = \sigma^2$. Thus the expected value of the sample variance is equal to the true population variance.

6.5 Some properties of estimators

In many situations it is possible to find several statistics which could be used to estimate some unknown parameter. For example if a random sample of n observations is taken from $N(\mu, \sigma^2)$ then three possible point estimates of μ are the sample mean, the sample median and the average of the largest and smallest observations. In order to decide which of them are 'good' estimates, we have to look at the properties of the sampling distributions of the different statistics. The statistic is called an estimator and is a random variable which varies from sample to sample. The word estimate, as opposed to estimator, is usually reserved for a particular value of the statistic.

One desirable property for an estimator is that of unbiasedness. An estimator is said to be *unbiased* or accurate if the mean of its sampling distribution is equal to the unknown parameter. Denote the unknown parameter by μ and the estimator by $\hat{\mu}$. (The ˆ or 'hat' over μ is the usual way of denoting an estimator). Thus $\hat{\mu}$ is an unbiased estimator of μ if

$$E(\hat{\mu}) = \mu.$$

For example the sample mean, $\bar{x}$, is an unbiased estimator of the population mean μ, because by Theorem 1 on p. 112 we have

$$E(\bar{x}) = \mu.$$

Unbiasedness by itself is not enough to ensure that an estimator is 'good'. In addition we would like the sampling distribution to be clustered closely round the true value. Thus given two unbiased estimators, $\hat{\mu}_1$ and $\hat{\mu}_2$, we would choose the one with the smaller variance or standard error. This brings in the idea of *efficiency*. If $\text{var}(\hat{\mu}_1) > \text{var}(\hat{\mu}_2)$ and $\hat{\mu}_1$, $\hat{\mu}_2$ are both unbiased estimators, then the relative efficiency of $\hat{\mu}_1$ with respect to $\hat{\mu}_2$ is given by

$$\frac{\text{var}(\hat{\mu}_2)}{\text{var}(\hat{\mu}_1)},$$

which is a number between 0 and 1. The unbiased estimator whose standard error is the smallest possible is sometimes called the *minimum variance unbiased estimator*.

But it is important to realize that there are many good estimators which are not unbiased. An estimator with a small bias and a small standard error may be better than an unbiased estimator with a large standard error. The idea of efficiency can be extended to consider biased estimators by calculating the expected mean square error, $E(\hat{\mu} - \mu)^2$. This is equal to the variance of $\hat{\mu}$ only when $\hat{\mu}$ is unbiased. The relative efficiency of $\hat{\mu}_1$ with respect to $\hat{\mu}_2$ is given quite generally by

$$\frac{E(\hat{\mu}_2 - \mu)^2}{E(\hat{\mu}_1 - \mu)^2}.$$

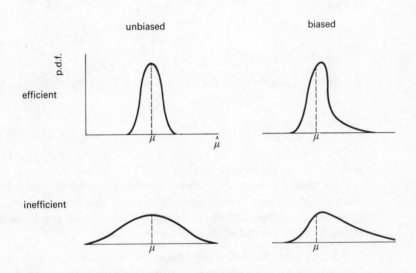

Figure 34   Types of estimators

Four types of sampling distribution are shown in Figure 34. An unbiased efficient estimator is clearly preferable to a biased inefficient estimator.

One further desirable property of a good estimator is that of consistency; which, roughly speaking, says that the larger the sample size n the closer the statistic will be to the true value. An estimator, $\hat{\mu}$, is said to be *consistent* if

(a) $E(\hat{\mu}) \rightarrow \mu$ as $n \rightarrow \infty$

(b) $\text{var}(\hat{\mu}) \rightarrow 0$ as $n \rightarrow \infty$

Property (a) clearly holds for unbiased estimators. Moreover the sample mean, for example, is consistent as $\text{var}(\bar{x}) = \sigma^2/n$ tends to zero as $n \to \infty$.

In some situations there is an estimator which is 'best' on all counts. But in other situations the best unbiased estimator may have a larger mean square error than some biased estimator and then the choice of the estimator to be used depends on practical considerations.

Example 4

If a random sample of n observations is taken from $N(\mu, \sigma^2)$, two possible estimates of μ are the sample mean and the sample median. We know

$$\text{var}(\bar{x}) = \frac{\sigma^2}{n}.$$

It can also be shown that the median is unbiased and that

$$\text{var(median)} = \frac{\pi\sigma^2}{2n}.$$

Thus the efficiency of the median is $2/\pi$ or about 64 per cent. Thus it is better to use the sample mean.

Note that for skewed distributions the median will give a biased estimate of the population mean. Moreover this bias does not tend to zero as the sample size increases and so the median is not a consistent estimate of the mean for skewed distributions.

We now state the following important results. If a random sample, size n, is taken from a normal distribution with mean μ and variance σ^2 then

(1) $\bar{x}$ *is the minimum variance unbiased estimate of* μ,

(2) s^2 *is the minimum variance unbiased estimate of* σ^2.

We have already proved that $E(\bar{x}) = \mu$, so $\bar{x}$ is an unbiased estimate of μ. Similarly it can be shown that

$$E(s^2) = E\left\{\frac{\sum(x_i - \bar{x})^2}{n-1}\right\} = \sigma^2,$$

so that s^2 is an unbiased estimate of σ^2. This is the reason that we chose this formula for s^2 in Chapter 2. If we put n instead of $(n-1)$

in the denominator of the formula for s^2, then we will get a biased estimate of σ^2. This fact, which puzzles many students, arises because the sample values are 'closer' on average to $\bar{x}$ than they are to μ. However, in the rare case where the population mean μ is known, but the variance σ^2 is not, the unbiased estimate of σ^2 would indeed be $\sum(x_i - \mu)^2/n$, with n in the denominator. The denominator in the formula for s^2 is called the number of degrees of freedom of the estimate. When μ is unknown and the observations are compared with $\bar{x}$, there is one linear constraint on the values of $(x_i - \bar{x})$, since $\sum(x_i - \bar{x}) = 0$, and so one degree of freedom is 'lost'.

More generally in Chapter 10 we will see that the number of degrees of freedom can be thought of as the number of independent comparisons available. With a random sample size n, each observation can be independently compared with the $(n-1)$ other observations and so there are $(n-1)$d.f. This is the same as the number of degrees of freedom of the estimate $\sum(x_i - \bar{x})^2/(n-1)$. If in addition we know the true mean μ, then each of the n observations can be independently compared with μ and so there are n d.f. This is the same as the number of degrees of freedom of the estimate $\sum(x_i - \mu)^2/n$.

The fact that $\bar{x}$ and s^2 are minimum variance estimates of μ and σ^2 requires a mathematical proof which is beyond the scope of this book.

6.6 General methods of point estimation

We have seen that the mean of a sample from a normal distribution is the best estimate of the population mean. Fortunately there are many cases where such intuitively obvious estimates are indeed the best. However, other situations exist where it is not obvious how to find a good estimate of an unknown parameter. Thus we will describe two general methods of finding point estimates.

6.6.1 *The method of moments*

Suppose that n observations, $x_1, x_2, \ldots, x_n$, are made on a random variable, X, whose distribution depends on one or more unknown parameters. The kth sample moment of the data is defined to be

$$m'_k = \sum_{i=1}^{n} \frac{x_i^k}{n}.$$

The kth population moment is defined to be

$$\mu'_k = E(X^k)$$

and this will depend on the unknown parameters.

The method of moments consists of equating the first few sample moments with the corresponding population moments, to obtain as many equations as there are unknown parameters. These can then be solved to obtain the required estimates. This procedure usually gives fairly simple estimates which are consistent. However they are sometimes biased and sometimes rather inefficient.

Example 5

Suppose we suspect that a variable has a distribution with p.d.f.

$$f(x) = \begin{cases} (k+1)x^k & (0 \leqslant x \leqslant 1, \quad k > 0), \\ 0 & \text{(otherwise)}. \end{cases}$$

and that the following values are observed.

0·2   0·4   0·5   0·7   0·8   0·8   0·9   0·9.

Estimate the unknown parameter k.

As there is only one unknown parameter, the method of moments consists simply of equating the sample mean with the population mean.

The sample mean $= m'_1 = \dfrac{5\cdot 2}{8} = 0.65$.

The population mean $= \mu'_1 = \displaystyle\int_0^1 (k+1)x^k x \, dx$

$$= \frac{(k+1)}{(k+2)}.$$

Thus the estimate of k is given by

$$\frac{(\hat{k}+1)}{(\hat{k}+2)} = 0.65$$

giving

$$\hat{k} = 0.86.$$

6.6.2 The method of maximum likelihood

This second method of estimation gives estimates which, besides being consistent, have the valuable property that, for large n, they are the most efficient. However, the estimates may be biased as may those obtained by the method of moments.

We introduce the idea of the likelihood function by means of an example. Suppose that we want to estimate the binomial parameter p, the probability of a success in a single trial, and that ten trials are performed resulting in seven successes. We know that the probability of observing x successes in ten trials is $^{10}C_x p^x(1-p)^{10-x}$. This is a probability function in which p is fixed and x varies between 0 and 10. However, in this particular experiment we observe $x = 7$ and would like to estimate p. If the value $x = 7$ is inserted in the probability function, it can now be thought of as a function of p in which case it is called a *likelihood function*. We write

$$L(p) = {}^{10}C_7 p^7(1-p)^3.$$

The value of p, say $\hat{p}$, which maximizes this function, is called the maximum likelihood estimate. Thus the method of maximum likelihood selects the value of the unknown parameter for which the probability of obtaining the observed data is a maximum. In other words the method selects the 'most likely' value for p. The true value of p is of course fixed and so it may be a trifle misleading to think of p as a variable in the likelihood function. This difficulty can be avoided by calling the likelihood a function of the unknown parameter.

In order to maximize the likelihood function it is convenient to take logs, thus giving

$$\log_e L(p) = \log_e {}^{10}C_7 + 7 \log_e p + 3 \log_e(1-p)$$

$$\frac{d \log L(p)}{dp} = \frac{7}{p} - \frac{3}{1-p}.$$

The likelihood function is maximized when $[d \log L(p)]/dp$ is zero. Thus we have

$$\frac{7}{\hat{p}} - \frac{3}{(1-\hat{p})} = 0$$

giving

$$\hat{p} = \tfrac{7}{10}.$$

Thus the intuitive estimate is also the maximum likelihood estimate.

More generally with a series of observations, $x_1, x_2, \ldots, x_n$, the likelihood function is obtained by writing down the joint probability of observing these values in terms of the unknown parameter. With a continuous distribution, the likelihood function is obtained by writing down the joint probability density function.

The method of maximum likelihood can also be used when the distribution depends on more than one parameter. For example, the normal distribution depends on two parameters, μ and σ^2, so that the likelihood is a function of two unknown parameters. To obtain the maximum value of this function we differentiate partially, first with respect to μ, and then with respect to σ; equate the derivatives to zero and solve to obtain the maximum likelihood estimates. With a sample size n, the maximum likelihood estimates turn out to be

$$\hat{\mu} = \bar{x}$$

$$\hat{\sigma}^2 = \frac{\sum(x_i - \bar{x})^2}{n}.$$

Notice that the estimate of σ^2 is biased, and so the usual estimate

$$s^2 = \frac{\sum(x_i - \bar{x})^2}{n-1}$$

is preferred for small samples. (Actually s^2 is the *marginal* likelihood estimate of σ^2, but this method of estimation will not be discussed in this book.)

Example 6

The Poisson distribution is given by

$$P_r = \frac{e^{-\lambda}\lambda^r}{r!} \qquad (r = 0, 1, 2, \ldots).$$

A sample of n observations, $x_1, x_2, \ldots, x_n$, is taken from this distribution. Derive the maximum likelihood estimate of the Poisson parameter λ.

The joint probability of observing $x_1, x_2, \ldots, x_n$ for a particular value of λ is given by

$$\prod_{i=1}^{n} \frac{e^{-\lambda}\lambda^{x_i}}{x_i!}.$$

Thus the likelihood function is given by

$$L(\lambda) = \prod_{i=1}^{n} \frac{e^{-\lambda}\lambda^{x_i}}{x_i!},$$

where the observations are known and λ is unknown.

$$\log_e L(\lambda) = (-\lambda + x_1 \log_e \lambda - \log_e x_1!) + (-\lambda + x_2 \log_e \lambda - \log_e x_2!) +$$

$$+ \ldots + (-\lambda + x_n \log_e \lambda - \log_e x_n!)$$

$$= -n\lambda + \log_e \lambda \sum x_i - \sum \log_e x_i!$$

$$\frac{d \log_e L(\lambda)}{d\lambda} = -n + \frac{\sum x_i}{\lambda}.$$

When this is zero we have

$$\hat{\lambda} = \frac{\sum x_i}{n} = \bar{x}.$$

Thus the intuitive estimate, $\bar{x}$, is also the maximum likelihood estimate of λ. (It is also the *method of moments* estimate.)

Example 7

Find the maximum likelihood estimate for the problem in Example 4.
    For a particular value of k, the joint probability density function of observing a series of values $x_1, x_2, \ldots, x_8$ is given by

$$(k+1)x_1^k \times (k+1)x_2^k \times \ldots \times (k+1)x_8^k.$$

Thus the likelihood function is given by

$$L(k) = (k+1)^8 (x_1)^k (x_2)^k \ldots (x_8)^k$$

$$\log_e L(k) = 8 \log_e(k+1) + k \sum_{i=1}^{8} \log_e x_i$$

$$\frac{d \log_e L(k)}{dk} = \frac{8}{k+1} + \sum_{i=1}^{8} \log_e x_i.$$

When this is zero we have

$$\hat{k}+1 = -\frac{8}{\sum\limits_{i=1}^{8} \log_e x_i}$$

$$= \frac{-8}{-4\cdot24}$$

$$= 1\cdot89$$

$$\hat{k} = 0\cdot89.$$

This estimate is very close to the estimate obtained by the method of moments.

6.7 Interval estimation

We have described two general methods of deriving point estimates of an unknown parameter. However it is usually preferable to find an interval estimate. This estimate is usually constructed in such a way that we have a certain confidence that the interval does contain the unknown parameter. The interval estimate is then called a *confidence interval*. We will concentrate on the important problem of finding a confidence interval for the population mean.

6.7.1 *Confidence interval for μ with σ known*

Suppose a random sample, size n, is taken from a distribution with unknown mean μ but whose standard deviation σ is known. This situation is somewhat uncommon but is a useful introduction to the case where σ is also unknown.

If the distribution is normal we have seen that $\bar{x}$ is $N(\mu, \sigma^2/n)$, so

$$z = \frac{\bar{x}-\mu}{\sigma/\sqrt{n}} \quad \text{is} \quad N(0, 1).$$

Now 95 per cent of the standard normal distribution lies between $\pm1\cdot96$. Therefore

$$\text{probability}\left(-1\cdot96 < \frac{\bar{x}-\mu}{\sigma/\sqrt{n}} < +1\cdot96\right) = 0\cdot95.$$

But

$$\frac{\bar{x}-\mu}{\sigma/\sqrt{n}} < 1\cdot96 \quad \text{implies} \quad \bar{x}-1\cdot96\frac{\sigma}{\sqrt{n}} < \mu$$

and

$$-1{\cdot}96 < \frac{\bar{x}-\mu}{\sigma/\sqrt{n}} \quad \text{implies} \quad \mu < \bar{x}+1{\cdot}96\frac{\sigma}{\sqrt{n}}.$$

Thus the expression can be rearranged to give

$$\text{probability}\left(\bar{x}-1{\cdot}96\frac{\sigma}{\sqrt{n}} < \mu < \bar{x}+1{\cdot}96\frac{\sigma}{\sqrt{n}}\right) = 0{\cdot}95.$$

The interval between $\bar{x}-1{\cdot}96\,\sigma/\sqrt{n}$ and $\bar{x}+1{\cdot}96\,\sigma/\sqrt{n}$ is called the 95 per cent confidence interval for μ. In other words given the sample mean $\bar{x}$, we are 95 per cent confident that this interval will contain μ. The two endpoints of the confidence interval are called the *confidence limits* for the unknown parameter.

The meaning of the above probability statement must be clearly understood. If $\bar{x}$ is a random variable then this statement is certainly true. In practice only one value is available and the confidence interval either will or will not contain μ, in which case the probabilities are one or nought respectively. Then the above statement must be interpreted as follows: if for each experiment like this we were to claim that μ lay within such a confidence interval, then 95 per cent of these claims would be true in the long run. For one particular experiment the probability or confidence of 95 per cent expresses the odds at which we would be prepared to bet that the confidence interval does contain μ.

Because of the central limit theorem a similar confidence interval is obtained when the sample is taken from a non-normal distribution provided that a reasonably large sample is taken.

Example 8

If the sample mean, $\bar{w}$, of the twelve measurements in Example 1 were found to be 12·91 per cent, give a 95 per cent confidence interval for the true percentage w.

The 95 per cent confidence interval for w is given by

$$\bar{w}\pm 1{\cdot}96\frac{\sigma}{\sqrt{n}} = 12{\cdot}91\pm 1{\cdot}96\times 0{\cdot}58$$

$$= 12{\cdot}91\pm 1{\cdot}14.$$

By a similar argument we can obtain the confidence intervals corresponding to stronger or weaker degrees of confidence. For example, 99 per cent of the normal distribution lies within 2·58 standard deviations from the mean. Thus the 99 per cent confidence interval is given by

$$\bar{x} \pm 2 \cdot 58 \frac{\sigma}{\sqrt{n}}.$$

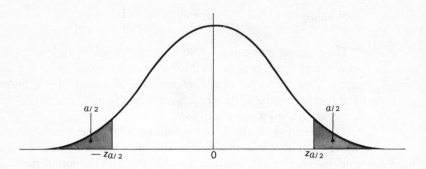

Figure 35

Generally to obtain the $100(1-\alpha)$ per cent confidence interval, we find the value, $z_{\frac{1}{2}\alpha}$, such that a proportion $\frac{1}{2}\alpha$ of the standard normal distribution lies above it. By symmetry a proportion $\frac{1}{2}\alpha$ lies below $-z_{\frac{1}{2}\alpha}$ so that the $100(1-\alpha)$ per cent confidence interval is given by

$$\bar{x} \pm z_{\frac{1}{2}\alpha} \frac{\sigma}{\sqrt{n}}.$$

6.7.2 Confidence interval for μ with σ unknown

In practice we usually find that the true standard deviation, σ, is unknown and that the sample standard deviation, s, is used to estimate σ. Then if a random sample size n is drawn, an estimate of the standard error of the sample mean $\bar{x}$ is given by $s/\sqrt{n}$.

Then we might expect the 95 per cent confidence interval for μ to be of the form $\bar{x} + 1\cdot 96\, s/\sqrt{n}$, simply replacing σ by s in our previous formula. For large samples ($n > 30$) we can in fact get a good approximation with this formula. However for small samples ($n < 30$) we must take a rather wider interval, since s is no longer such a good estimate of σ, and there will be appreciable variation in s from sample to sample.

If we denote the 95 per cent confidence interval for μ, for a sample size n, by

$$\bar{x} \pm t_c \frac{s}{\sqrt{n}}, \qquad\qquad \textbf{6.8}$$

then some typical values for t_c are given in Table 10.

Table 10
Values of t_c for a
95 per cent Confidence Interval

n	t_c
4	3·18
8	2·36
12	2·20
20	2·09
∞	1·96

Thus although t_c increases quite rapidly for very small n, a useful approximation for $n > 25$, which is worth remembering, is that the 95 per cent confidence interval is $\bar{x} \pm 2$(estimated standard error of sample mean).

Example 9

A sample of eight observations is taken from a distribution with unknown mean μ. The sample mean $\bar{x}$ is 7·91 and the sample standard deviation s is 0·67. Thus an estimate of the standard error of the sample mean is $0·67/\sqrt{8} = 0·237$. From Table 10 we find that for a sample size 8 the 95 per cent confidence interval for the true mean μ is given by $\bar{x} \pm 2·36 \times s/\sqrt{n}$.

Thus the 95 per cent confidence interval is given by $7·91 \pm 0·56$.

In equation **6.8** t_c is actually the percentage point of a distribution called the t-distribution. We have already seen that if random samples, size n, are drawn from a normal distribution, mean μ and variance σ^2, then the sampling distribution of the random variable

$$z = \frac{\bar{x} - \mu}{\sigma/\sqrt{n}} \text{ is } N(0, 1).$$

However, if σ is replaced with its sample estimate s, then the random variable

$$t = \frac{\bar{x} - \mu}{s/\sqrt{n}}$$

has a sampling distribution which is somewhat more spread out than the standard normal distribution. It can be shown that the statistic t follows a distribution called the t-distribution, which is related to the normal and χ^2 distributions (see Appendix A). This distribution depends on a parameter called the number of degrees of freedom, which is the same as the number of degrees of freedom of the estimate s. In the above situation, where s^2 was calculated from samples size n, there are $(n-1)$ degrees of freedom. An example of a t-distribution, with three degrees of freedom, is given in Figure 36 together with a standard normal distribution for comparison.

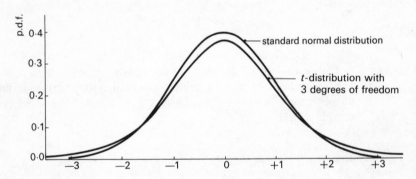

Figure 36

The t-distribution is sometimes called the Student t-distribution after W. S. Gossett who studied the distribution and published papers on it under the pen name 'Student'. Like the standard normal distribution it is symmetric with mean zero. The percentage point, $t_{\alpha,\nu}$, is chosen so that a proportion α of the t-distribution, with ν d.f., lies above it. These percentage points are tabulated in Table 2, Appendix B.

The percentage points which are required to establish confidence intervals for μ are such that

$$P(|t| > t_c) = \alpha.$$

Thus $t_c = t_{\frac{1}{2}\alpha}$ with the appropriate number of degrees of freedom (see Figure 37).

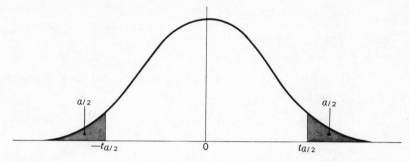

Figure 37

We are now in a position to find a confidence interval for μ for any degree of confidence we wish. If $\bar{x}$ and s^2 are calculated from a sample size n, the $100(1-\alpha)$ per cent confidence interval for μ is given by

$$\bar{x} \pm t_{\frac{1}{2}\alpha, n-1} \frac{s}{\sqrt{n}}.$$

Example 10

Use the data given in Example 9 to establish a 99 per cent confidence interval for the true mean μ.

There are eight observations, so that the sample estimate s is based on $(8-1)$ d.f. From Table 2, Appendix B, we find

$$t_{0.005, 7} = 3.50.$$

From Example 9 we have $s/\sqrt{n} = 0.237$. Thus the 99 per cent confidence interval for μ is given by

$$7.91 \pm 3.50 \times 0.237 = 7.91 \pm 0.83.$$

This is about half as wide again as the 95 per cent confidence interval for μ.

6.7.3 *Confidence interval for σ^2*

Interval estimates for other parameters can be derived in a similar way. For example we know that $(n-1)s^2/\sigma^2$ follows a χ^2 distribution with $(n-1)$ degrees of freedom. Thus

$$P\left(\chi^2_{1-\frac{1}{2}\alpha, n-1} < \frac{(n-1)s^2}{\sigma^2} < \chi^2_{\frac{1}{2}\alpha, n-1}\right) = \alpha.$$

This can be rearranged to give

$$P\left\{\frac{(n-1)s^2}{\chi^2_{\frac{1}{2}\alpha,n-1}} < \sigma^2 < \frac{(n-1)s^2}{\chi^2_{1-\frac{1}{2}\alpha,n-1}}\right\} = \alpha.$$

Thus the $100(1-\alpha)$ per cent confidence interval for σ^2 lies between

$$\frac{(n-1)s^2}{\chi^2_{\frac{1}{2}\alpha,n-1}} \quad \text{and} \quad \frac{(n-1)s^2}{\chi^2_{1-\frac{1}{2}\alpha,n-1}}.$$

Exercises

1. A random sample is drawn from a population with a known standard deviation of 2·0. Find the standard error of the sample mean if the sample is of size (a) 9; (b) 25 and (c) 100. What sample size would give a standard error equal to 0·5?

2. If X, Y are independent random variables show that

variance $(X - Y) =$ variance $(X) +$ variance (Y).

Two random samples of size n_1 and n_2 are taken from a population with variance σ^2. The two sample means are $\bar{x}_1$ and $\bar{x}_2$. Show that the standard error of $\bar{x}_1 - \bar{x}_2$ is given by

$$\sigma\sqrt{\left(\frac{1}{n_1} + \frac{1}{n_2}\right)}.$$

3. A random sample size 10 is taken from a standard normal distribution. Find the values a and b such that

$$P(a < \bar{x}) = 0.95$$
$$P(\bar{x} < b) = 0.95$$

i.e.

$$P(a < \bar{x} < b) = 0.90.$$

Also use the χ^2 distribution to find the values c and d such that

$$P(c < s^2) = 0.95$$
$$P(s^2 < d) = 0.95$$

i.e.

$$P(c < s^2 < d) = 0.90.$$

Also find the values of a, b, c and d if the sample was taken from a normal distribution with a mean of six and a variance of four.

4. The probability density function of the exponential distribution is given by

$$f(x) = \lambda e^{-\lambda x} \qquad x > 0.$$

A random sample of n observations is taken from this distribution. Show that the method of moments estimate of λ is given by

$$\hat{\lambda} = \frac{1}{\bar{x}} \quad \text{where } \bar{x} \text{ is the sample mean.}$$

Also show that the maximum likelihood estimate of λ is the same as the method of moments estimate.

5. The percentage of copper in a certain chemical is measured six times. The standard deviation of repeated measurements is known to be 2·5; the sample mean is 14·1. Give a 95 per cent confidence interval for the true percentage of copper, assuming that the observations are approximately normally distributed.

6. The percentage of copper in a certain chemical is measured six times. The sample mean and sample standard deviation are found to be 14·1 and 2·1 respectively. Give a 95 per cent confidence interval for the true percentage of copper, assuming that the observations are approximately normally distributed.

7. A random sample, size n_1, is taken from $N(\mu_1, \sigma^2)$ and a second random sample, size n_2, is taken from $N(\mu_2, \sigma^2)$. If the sample means are denoted by $\bar{x}_1$, $\bar{x}_2$ it can be shown that

$$\frac{(\bar{x}_1 - \bar{x}_2) - (\mu_1 - \mu_2)}{\sigma\sqrt{(1/n_1 + 1/n_2)}} \quad \text{is } N(0, 1).$$

Show that the $100(1 - \alpha)$ per cent confidence interval for $(\mu_1 - \mu_2)$, assuming σ is known, is given by

$$(\bar{x}_1 - \bar{x}_2) \pm z_{\frac{1}{2}\alpha}\sigma\sqrt{\left[\frac{1}{n_1} + \frac{1}{n_2}\right]}.$$

References

HOEL, P. G. (1964), *Introduction to Mathematical Statistics*, Wiley, 3rd edn.
KENDALL, M. G., and BABINGTON-SMITH, B. (1951), *Tables of Random Sampling Numbers*, Cambridge University Press.
FISHER, R. A., and YATES, F. (1963), *Statistical Tables for Biological, Agricultural and Medical Research*, Oliver & Boyd, 6th edn.
THE RAND CORPORATION (1955), *A Million Random Digits*, Free Press.

Chapter 7
Significance tests

7.1 Introduction

We have seen that statistics is concerned with making deductions from a sample of data about the population from which the sample is drawn. In Chapter 6 we considered the problem of estimating the unknown parameters of the population. In this chapter we shall see how to carry out a significance test in order to test some theory about the population.

The commonest situation is when the population can be described by a probability distribution which depends on a single unknown parameter μ. On the basis of the experimental results we might wish, for example, to accept or reject the theory that μ has a particular value μ_0. A numerical method for testing such a theory or hypothesis is called a *significance test*.

Example 1

One of the commonest problems facing the engineer is that of trying to improve an industrial process in some respect. In particular he may wish to compare a new process with an existing process. However, because of changeover costs, he will need to be reasonably certain that the new process really is better before making the necessary changes.

As an example let us suppose that the strength of steel wire made by an existing process is normally distributed with mean $\mu_0 = 1250$ and standard deviation $\sigma = 150$. A batch of wire is made by a new process, and a sample of 25 measurements gives an average strength of $\bar{x} = 1312$. (We assume that the standard deviation of the measurements does not change.) Then the engineer must decide if the difference between $\bar{x}$ and μ_0 is strong enough evidence to justify changing to the new process. The method of testing the data to see if the results are considerably or significantly better is called a significance test.

In any such situation the experimenter has to weigh the evidence and, if possible, decide between two rival possibilities. The hypothesis which we want to test is called the *null hypothesis*, and is denoted by H_0. Any other hypothesis is called the *alternative hypothesis*, and is denoted by H_1.

In the case of the engineer in Example 1, he must decide whether to accept or reject the hypothesis that there is no difference between the new and existing processes. This theory is known as the null hypothesis because it assumes there is no difference between the two processes.

Let us denote the mean strength of wire produced by the new process by μ, so that $\bar{x}$ is the sample estimate of μ. Then the null hypothesis is given by

$$H_0 : \mu = \mu_0.$$

The second possible theory in this situation is that the new process is better than the existing process and this is the alternative hypothesis. It is also useful to formulate H_1 precisely as follows:

$$H_1 : \mu > \mu_0.$$

At first the student may have difficulty in deciding which of two theories is the null hypothesis and which is the alternative hypothesis. The important point to remember is that the null hypothesis must be assumed to be true until the data indicates otherwise, in much the same way that a prisoner is assumed innocent until proved guilty. Thus the burden of proof is on H_1 and the experimenter is interested in departures from H_0 rather than from H_1.

Example 2

If a new industrial process is compared with an existing process, the choice of the null hypothesis depends on a number of considerations and several possibilities arise.

(a) If the existing process is reliable and changeover costs are high, the burden of proof is on the new process to show that it really is better. Then we choose

      H_0: existing process is as good as or better than the new process.

and H_1: new process is an improvement.

This is the situation discussed in Example 1.

(b) If the existing process is unreliable and changeover costs are low, the burden of proof is on the existing process. Then we choose

H_0: new process is as good as or better than the existing process. and H_1: existing process is better than the new process.

In these two extreme cases there is no difficulty in choosing the correct null hypothesis. But in practice, the situation may be somewhere in between. Then, in addition to choosing the more natural null hypothesis, it is also essential to make a careful choice of the sample size, as described in section 7.8, in order to ensure that the risk of making a wrong decision is acceptably small. Alternatively a decision theory approach may be possible by allocating costs to the different eventualities, but this approach will not be described in this book.

Example 3

A new drug is tested to see if it is effective in curing a certain disease. The Thalidomide tragedy has emphasized that a drug must not be put on the market until it has been rigorously tested. Thus we must assume that the drug is not effective, or is actually harmful, until the tests indicate otherwise. The null hypothesis is that the drug is not effective. The alternative hypothesis is that the drug is effective.

A second factor to bear in mind when choosing the null hypothesis is that it should nearly always be precise, or be easily reduced to a precise hypothesis. For example when testing

$$H_0: \mu \leqslant \mu_0$$
against $$H_1: \mu > \mu_0,$$

the null hypothesis does not specify the value of μ exactly and so is not precise. But in practice we would proceed as if we were testing

$$H_0: \mu = \mu_0$$
against $$H_1: \mu > \mu_0$$

and here the null hypothesis is precise.

Further comments on choosing a precise null hypothesis are included at the end of section 7.8.

7.1.1 *Test statistic*

Having decided on the null and alternative hypotheses, the next step is to calculate a statistic which will show up any departure from the null hypothesis.

In the case of Example 1, common sense suggests that the larger the difference between $\bar{x}$ and μ_0 the more likely it is that the new process has increased the strength. If we divide this difference by the standard error of $\bar{x}$, which is $\sigma/\sqrt{25}$, we have the test statistic

$$z = \frac{\bar{x} - \mu_0}{\sigma/\sqrt{25}}.$$

However we must remember that even if there is no difference between the two processes (that is, H_0 is true), we cannot expect $\bar{x}$ to be exactly equal to μ_0. If H_0 is true, we have seen in section 6.3 that $\bar{x}$ will have a sampling distribution mean μ_0 and standard error $\sigma/\sqrt{25}$. Thus z will follow a standard normal distribution, $N(0, 1)$. Thus if H_0 is true we are unlikely to get a value of z bigger than about $+2$ so that if $(\bar{x} - \mu_0)/(\sigma/\sqrt{25})$ is larger than this, then doubt is thrown on the null hypothesis.

7.1.2 *Level of significance*

This is the probability of getting a result which is as extreme, or more extreme, than the one obtained. A result which is unlikely to occur if H_0 is true (that is, which has a low level of significance) is called a *significant result*.

Using the data of Example 1, we first calculate the observed value of the test statistic

$$z_0 = \frac{1312 - 1250}{150/\sqrt{25}} = 2 \cdot 06.$$

From normal tables we find

probability $(z > 2 \cdot 06) = 1 - F(2 \cdot 06)$

$$= 0 \cdot 0197.$$

Thus there is about a 2 per cent chance of getting a more extreme result. As this result is so unlikely, we are inclined to reject the null hypothesis and accept the alternative hypothesis that there has been an improvement.

7.1.3   The interpretation of a significant result

It is important to realize from the outset that we can rarely prove with absolute certainty that H_0 is or is not true. If the level of significance of the test statistic is fairly low, then doubt is thrown on the null hypothesis and we will be inclined to reject it. Nevertheless there is still a small possibility that the observed data could have resulted from the given null hypothesis.

If the level of significance of the test statistic is less than 5 per cent, we say that the result is significant at the 5 per cent level. This is generally taken to be reasonable evidence that H_0 is untrue. If the result is significant at the one per cent level of significance, it is generally taken to be fairly conclusive evidence that H_0 is untrue.

On the other hand if the level of significance is fairly high (> 5 per cent) then this simply shows that the data is quite consistent with the null hypothesis. This does not mean that H_0 is definitely true, as the sample size may not have been large enough to spot a fairly small departure from H_0. But, for the time being at least, we have no evidence that H_0 is untrue and so will accept it.

Thus if we had obtained the result $\bar{x} = 1290$ in Example 1, the level of significance would be greater than 5 per cent and we would not have enough evidence to reject H_0, as such a result is quite likely to occur if H_0 is true. But, if the difference of forty between $\bar{x}$ and μ_0 is judged to be a worthwhile improvement, then a larger sample should be taken in order to try to detect a smaller improvement than was possible with a sample of twenty-five. Where possible the sample size should be chosen beforehand so that the smallest significant difference, which is about two standard errors $(2\sigma/\sqrt{n})$, is much less than the smallest improvement considered to be of practical importance (see section 7.8 and Exercise 4).

At this point it is worth emphasizing the distinction between statistical and practical significance. For example, suppose that the result $\bar{x} = 1270$ was obtained in Example 1, and that the engineer decided that the difference of twenty between this value and μ_0 was not sufficiently large to justify changing to the new process. In other words, the difference of twenty was judged not to be of practical significance. Then there is no point in doing a significance test to see if the result is statistically significant, since the engineer's actions will not be affected by the result of such a test. Thus we must remember that a result may be statistically significant but not practically significant.

7.1.4 One-tailed and two-tailed tests

In Example 1 we were only interested in values of $\bar{x}$ significantly higher than μ_0. Any such test which only takes account of departures from the null hypothesis in one direction is called a *one-tailed* test (or *one-sided* test).

However, other situations exist in which departures from H_0 in two directions are of interest. For example, suppose that a chemical process has optimum temperature T and that measurements are made at regular intervals. The chemist is interested in detecting a significant increase *or* decrease in the observed temperature and so a *two-tailed* test is appropriate. It is important that the scientist should decide if a one-tailed or two-tailed test is required before the observations are taken.

Example 4

According to a certain chemical theory, the percentage of iron in a certain compound should be 12·1. In order to test this theory it was decided to analyse nine different samples of the compound to see if the measurements would differ significantly from 12·1 per cent.

Before carrying out the analyses, we must specify the null and alternative hypotheses and also decide if a one-tailed or two-tailed test is appropriate.

Denote the unknown mean percentage of iron by μ. Then the null hypothesis is given by

$H_0 : \mu = 12 \cdot 1 \%.$

The alternative hypothesis is given by

$H_1 : \mu \neq 12 \cdot 1 \%.$

As we are interested in significantly high *or* low results, a two-tailed test is appropriate. The actual measurements and the resulting analysis will be given later in the chapter.

For tests based on the normal distribution, the level of significance of a result in a two-tailed test can be obtained by doubling the level of significance which would be obtained if a one-tailed test was carried out on the same result (see section 7.2).

7.1.5 Critical values

We have seen that the lower the level of significance of a particular result, the less likely it is that the null hypothesis is true. If the observed level of significance is very small, then a decision can be made to reject the null hypothesis. If very large then a decision can be made to accept the null hypothesis. But if the observed level of significance is around 5 per cent, the results are rather inconclusive and the experimenter may decide to take some more measurements.

However, if a definite decision has to be taken, the experimenter should choose a particular level of significance and reject the null hypothesis if the observed level of significance is less than this. The 5 per cent and the 1 per cent levels are commonly used. This value should be chosen before the observations are taken. For example, if the 5 per cent level is chosen and the observed level of significance is less than this, then we say the result is significant at the 5 per cent level.

A *critical value* of the test statistic can be calculated which will correspond to the chosen significance level. Thus in Example 1 we could choose to reject H_0 if the observed level of significance is less than 5 per cent. But probability $(z > 1.64) = 0.05$. Thus the critical value of the test statistic, z, is 1.64. If the observed value of z is greater than 1.64, then H_0 must be rejected.

A variety of significance tests exist each of which is appropriate to a particular situation. In the remainder of this chapter we will describe some of the more important tests.

7.2 Tests on a sample mean

In this section we assume that a random sample, size n, is taken from a normal distribution with unknown mean μ. The sample mean is denoted by $\bar{x}$. We are interested in a particular value μ_0 for μ, and ask the question 'does $\bar{x}$ differ significantly from μ_0?'. Thus the null hypothesis is given by

$$H_0 : \mu = \mu_0.$$

The method of analysis depends on whether or not the population standard deviation σ is known.

7.2.1 σ known

This situation has been discussed in detail in Example 1. The test statistic is given by

$$z = \frac{\bar{x} - \mu_0}{\sigma / \sqrt{n}}.$$

If H_0 is true then z is a standard normal variable. Denote the observed value of z by z_0. If a two-tailed test is appropriate the level of significance is obtained by calculating

$$P(|z| \geqslant |z_0|) = 2 \times P(z \geqslant |z_0|)$$

from normal tables (see Figure 38).

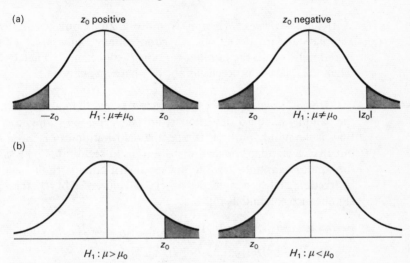

Figure 38   Shaded area gives level of significance
(a) two-tailed test H_1: $\mu \neq \mu_0$ (b) one-tailed test

If a one-tailed test is appropriate we find $P(z > z_0)$ if we are interested in significantly high values, or $P(z < z_0)$ if we are interested in significantly low values. Note that if z_0 is negative we have

$$P(z < z_0) = P(z > |z_0|).$$

by the symmetry of the normal distribution.

The 95 per cent and 99 per cent critical values for the test statistic z can be found from normal tables and are given in Table 11.

Table 11

	Two-tailed	One-tailed
95%	1·96	1·64
99%	2·58	2·33

7.2.2 σ unknown

When σ is unknown it is natural to replace it with the sample standard deviation, s, to obtain the test statistic

$$t = \frac{\bar{x} - \mu_0}{s/\sqrt{n}}.$$

(i) *Large samples*. When n is more than about twenty-five, the sample standard deviation is a good estimate of σ and, if the null hypothesis is true, the random variable t will be approximately $N(0, 1)$. Then the analysis proceeds as above where σ is known.

(ii) *Small samples – the t-test*. When n is less than about twenty-five the sample standard deviation is not such a good estimate of σ, so that, if the null hypothesis is true, the distribution of t is more spread out than a standard normal distribution. In section 6.7.2 we saw that the random variable t will follow a t-distribution with $(n-1)$ degrees of freedom. If we denote the observed value of t by t_0, the level of significance is found as follows:

two-tailed test $P(|t| \geqslant |t_0|) = 2P(t \geqslant |t_0|)$ (for $H_1 : \mu \neq \mu_0$),

one-tailed test $P(t \geqslant t_0)$ (for $H_1 : \mu > \mu_0$)
$P(t \leqslant t_0)$ (for $H_1 : \mu < \mu_0$).

These probabilities can be found from the table of percentage points of the t-distribution which is given in Appendix B. As described in section 6.7.2, the percentage point $t_{\alpha, \nu}$ is chosen so that there is a probability α of getting a larger observation from a t-distribution with ν degrees of freedom. For a two-tailed test if $|t_0|$ is larger than $t_{\alpha/2}$, then the result is significant at the 100α per cent significance level. For a one-tailed test, if t_0 is larger than t_α for $H_1 : \mu > \mu_0$ or if t_0 is less than $-t_\alpha$ for $H_1 : \mu < \mu_0$, then the result is significant at the 100α per cent significance level.

Example 4 continued

The analysis of the nine different samples of the compound gave the following results:

    11·7   12·2   10·9   11·4   11·3   12·0   11·1   10·7   11·6

Is the sample mean of these measurements significantly different from 12·1 per cent?

From the observations we compute

$$\bar{x} = 11 \cdot 43$$

$$s^2 = 0 \cdot 24$$

$$s = 0 \cdot 49.$$

The population standard deviation is unknown, so we compute the test statistic

$$t_0 = \frac{\bar{x} - \mu_0}{s/\sqrt{n}}$$

$$= \frac{11 \cdot 43 - 12 \cdot 1}{0 \cdot 49/\sqrt{9}}$$

$$= -4 \cdot 1.$$

As the sample standard deviation is computed from only nine measurements, we must use a t-test. If the null hypothesis is true the test statistic should follow a t-distribution with eight degrees of freedom.

From Table 2, Appendix B,

$$t_{0 \cdot 025, 8} = 2 \cdot 31.$$

As $|t_0|$ is larger than $t_{0 \cdot 025, 8}$, the probability of observing a result which is as extreme or more extreme than this is certainly less than $2 \times 0 \cdot 025 = 0 \cdot 05$. Thus the result is significant at the 5 per cent level. Moreover we have $t_{0 \cdot 005, 8} = 3 \cdot 36$ so that the result is also significant at the 1 per cent level. Thus we have fairly conclusive evidence that the percentage of iron is not 12·1 per cent.

7.3   Comparing two sample means

A common problem in statistics is that of comparing the means of two samples of data. Let us assume that a random sample, size n_1, is taken from a population having unknown mean μ_1 and that a second sample, size n_2, is taken from a population having unknown mean μ_2. The observations in the first sample will be denoted by $x_1, x_2, \ldots, x_{n_1}$, and in the second sample by $x_1', x_2', \ldots, x_{n_2}'$. The two sample means will be denoted by $\bar{x}_1$ and $\bar{x}_2$.

This section is concerned with answering the question 'does $\bar{x}_1$ differ significantly from $\bar{x}_2$?'. In other words the problem is to test the hypothesis $H_0 : \mu_1 = \mu_2$. The alternative hypothesis may be of the form $H_1 : \mu_1 \neq \mu_2$, $H_1 : \mu_1 > \mu_2$ or $H_1 : \mu_1 < \mu_2$. A two-tailed test is appropriate in the first case and a one-tailed test in the second and third.

The significance test proposed here depends on the following assumptions. Firstly, that both sets of observations are normally distributed; secondly, that the populations have the same variance σ^2; and thirdly, that the two samples are independent.

The method of analysis again depends on whether or not the population standard deviation σ is known.

7.3.1 σ known

Common sense suggests that the larger the difference between $\bar{x}_1$ and $\bar{x}_2$, the less likely H_0 is to be true. A standardized test statistic can be obtained by dividing $(\bar{x}_1 - \bar{x}_2)$ by its standard error, which can be found as follows. We know variance $(\bar{x}_1) = \sigma^2/n_1$ and variance $(\bar{x}_2) = \sigma^2/n_2$. In addition we know that the variance of the sum or difference of two independent random variables is equal to the sum of the variances of the two variables (see section 6.2). Thus

$$\text{variance } (\bar{x}_1 - \bar{x}_2) = \frac{\sigma^2}{n_1} + \frac{\sigma^2}{n_2}.$$

Thus the standard error of $(\bar{x}_1 - \bar{x}_2)$ is given by

$$\sqrt{(\sigma^2/n_1 + \sigma^2/n_2)} = \sigma\sqrt{(1/n_1 + 1/n_2)}.$$

The test statistic is given by

$$z = \frac{\bar{x}_1 - \bar{x}_2}{\sigma\sqrt{(1/n_1 + 1/n_2)}}.$$

If H_0 is true it can be shown that the random variable z is a standard normal variable. The level of significance of a particular result z_0 can be found as in section 7.2.

7.3.2 σ unknown

Following the method adopted in section 7.2 we replace σ with the sample standard deviation, s, to obtain the test statistic

$$t = \frac{\bar{x}_1 - \bar{x}_2}{s\sqrt{(1/n_1 + 1/n_2)}}.$$

The standard deviation of the first sample is given by

$$s_1^2 = \sum_{i=1}^{n_1} \frac{(x_i - \bar{x}_1)^2}{n_1 - 1}.$$

The standard deviation of the second sample is given by

$$s_2^2 = \sum_{i=1}^{n_2} \frac{(x_i' - \bar{x}_2)^2}{n_2 - 1}.$$

Then the combined unbiased estimate of σ^2 is given by

$$s^2 = \frac{(n_1 - 1)s_1^2 + (n_2 - 1)s_2^2}{n_1 + n_2 - 2}$$

$$= \frac{\sum(x_i - \bar{x}_1)^2 + \sum(x_i' - \bar{x}_2)^2}{n_1 + n_2 - 2}.$$

This result follows from the fact that if X, Y are independent χ^2 random variables with v_1, v_2 degrees of freedom then $X + Y$ is also a χ^2 random variable with $(v_1 + v_2)$ degrees of freedom. Thus

$$\left[\frac{(n_1 - 1)s_1^2}{\sigma^2} + \frac{(n_2 - 1)s_2^2}{\sigma^2}\right] \quad \text{is} \quad \chi^2_{(n_1 - 1) + (n_2 - 1)}$$

As the mean value of the χ^2 distribution is equal to the number of degrees of freedom we find

$$E\left[\frac{(n_1 - 1)s_1^2 + (n_2 - 1)s_2^2}{n_1 + n_2 - 2}\right] = \sigma^2$$

The denominator $(n_1 + n_2 - 2)$ represents the number of degrees of freedom of this estimate of the variance, and is equal to the number of degrees of freedom of the estimate s_1^2 plus the number of degrees of freedom of the estimate s_2^2. Note that if $n_1 = n_2$ then we have

$$s^2 = \frac{s_1^2 + s_2^2}{2}.$$

If the null hypothesis is true, it can be shown that the test statistic, t, follows a t-distribution with $(n_1 + n_2 - 2)$ degrees of freedom. Thus the level of significance of a particular result, t_0, can be found as before.

Example 5

Two batches of a certain chemical were delivered to a factory. For each batch ten determinations were made of the percentage of manganese in the chemical. The results were as follows:

Batch 1: 3·3 3·7 3·5 4·1 3·4 3·5 4·0 3·8 3·2 3·7
Batch 2: 3·2 3·6 3·1 3·4 3·0 3·4 2·8 3·1 3·3 3·6

Is there a significant difference between the two sample means? From the data we find

	Batch 1	Batch 2
$\sum x_i$	36·2	32·5
$\sum x_i^2$	131·82	106·23
$\sum x_i^2 - \dfrac{(\sum x_i)^2}{10}$	0·776	0·605

Hence

$$\bar{x}_1 = 3·62 \qquad \bar{x}_2 = 3·25$$

$$s_1^2 = \frac{0·776}{9} = 0·0862$$

$$s_2^2 = \frac{0·605}{9} = 0·0672.$$

It is reasonable to assume that both sets of observations are normally distributed and that both sample variances are estimates of the same population variance (see Example 10). The combined estimate of this variance is given by

$$s^2 = \frac{s_1^2 + s_2^2}{2} = 0·0767$$

$$s = 0·277.$$

If the true percentages of manganese in batch 1 and batch 2 are denoted by μ_1 and μ_2 respectively, the null hypothesis is given by $H_0: \mu_1 = \mu_2$, and the alternative hypothesis is given by $H_1: \mu_1 \neq \mu_2$. Thus a two-tailed test is appropriate. The test statistic is given by

$$t_0 = \frac{\bar{x}_1 - \bar{x}_2}{s\sqrt{(\frac{1}{10} + \frac{1}{10})}} = 2·98.$$

If H_0 is true the sampling distribution of t will be a t-distribution with eighteen degrees of freedom, as the estimate of s is based on eighteen degrees of freedom.

But from Table 2, Appendix B, we find $t_{0·005,18} = 2·88$, so that

probability $(|t| > 2·88) = 0·01$.

Thus the result is significant at the 1 per cent level and we have strong evidence that H_0 is untrue.

7.4 The t-test applied to paired comparisons

When comparing two different methods, it often happens that experiments are carried out in pairs. Then it is the *difference* between each pair of measurements which is of interest.

Example 6

In order to compare two methods for finding the percentage of iron in a compound, ten different compounds were analysed by both methods and the results are given below.

Compound number	Method A	B	Compound number	Method A	B
1	13·3	13·4	6	3·7	4·0
2	17·6	17·9	7	5·1	5·1
3	4·1	4·1	8	7·9	8·0
4	17·2	17·0	9	8·7	8·8
5	10·1	10·3	10	11·6	12·0

We ask the question 'is there a significant difference between the two methods of analysis?'

Note that it would be wrong to calculate the average percentage for method A and for method B and proceed as in section 7.3 because the variation between compounds will swamp any difference there may be between the two methods. Instead we compute the difference for each compound as below.

Compound number	Difference	Compound number	Difference
1	0·1	6	0·3
2	0·3	7	0·0
3	0·0	8	0·1
4	−0·2	9	0·1
5	0·2	10	0·4

If the two methods give similar results, the above differences should be a sample of ten observations from a population with mean zero.

Generally we have k pairs of measurements $x_{1j}, x_{2j}, (j = 1, 2, \ldots, k)$ which are independent observations from populations with means μ_{1j}, μ_{2j}. The null hypothesis is that each pair of means are equal.

$H_0 : \mu_{1j} = \mu_{2j}$ (for all j).

Then the differences

$$d_j = x_{1j} - x_{2j} \qquad (j = 1, \ldots, k)$$

will be a sample, size k, from a population with mean zero. Furthermore, if the populations are approximately normally distributed, the differences will also be approximately normally distributed. If the observed average difference is denoted by $\bar{d}$ and the standard deviation of the observed differences by s_d, then the standard error of $\bar{d}$ is given by $s_d/\sqrt{k}$. We now apply a t-test, as in section 7.2, by calculating the test statistic

$$t = \frac{\bar{d}}{s_d/\sqrt{k}}.$$

If H_0 is true, the distribution of t will be a t-distribution with $(k-1)$ degrees of freedom, as the estimate s_d is calculated from k differences.

Example 6 continued

$\bar{d}$ = average difference = 0·13

$$s_d^2 = \frac{\sum_{i=1}^{k} (d_i - \bar{d})^2}{k-1} = 0 \cdot 031 \qquad s_d = 0 \cdot 176$$

$$t_0 = \frac{\bar{d}}{s_d/\sqrt{10}} = 2 \cdot 33$$

The alternative hypothesis is of the form $H_1 : \mu_{1j} \neq \mu_{2j}$, for all j, and so a two-tailed test is required. From Table 2, Appendix B, we find

$$t_{0 \cdot 025, 9} = 2 \cdot 26.$$

Thus $P(|t| > 2 \cdot 26) = 0 \cdot 05$.

As t_0 is greater than 2·26, the result is significant at the 5 per cent level, and so we have reasonable evidence that H_0 is untrue.

7.5 The χ^2 goodness-of-fit test

We now turn our attention to a different type of problem. Data can often be classified into k mutually exclusive classes or categories. Then we need a test to see if the observed frequencies in each category are significantly different from those which could be expected if some hypothesis were true.

Let us suppose that this hypothesis suggests that $p_1, p_2, \ldots, p_k$ are the respective probabilities of the categories, where $\sum_{i=1}^{k} p_i = 1$. That is,

$H_0 : p_i$ = probability that ith outcome will occur $(i = 1, \ldots, k)$.

If n experiments are performed, the expected number of occurrences of the ith outcome is given by $e_i = np_i$. Denote the observed frequency of the ith outcome by o_i. Then we want to know if $o_1, \ldots, o_k$ are compatible with $e_1, \ldots, e_k$.

Theorem

If $o_1, \ldots, o_k$ and $np_1, \ldots, np_k$ are the observed and expected frequencies for the k possible outcomes of an experiment, then, for large n, the distribution of the quantity

$$\sum_{i=1}^{k} \frac{(o_i - np_i)^2}{np_i}$$

is approximately that of a χ^2 random variable with $(k-1)$ degrees of freedom. One degree of freedom is lost because of the constraint

$$\sum_{i=1}^{k} o_i = n = \sum_{i=1}^{k} e_i.$$

The χ^2 distribution was introduced in section 6.4 and percentage points are given in Table 3, Appendix B. The point $\chi^2_{\alpha, v}$ is such that there is a probability α of observing a larger value of χ^2, with v degrees of freedom. The observed value of the χ^2 test statistic will be denoted by χ^2_0. If the null hypothesis is not true, then we expect χ^2_0 to be 'large', and as we are only interested in significantly large values of the test statistic, we have a one-tailed test in which the level of significance of the observed result is given by $P(\chi^2 > \chi^2_0)$.

Example 7

Assuming that a die is fair we have

p_i = probability that an i turns up

$\quad = \frac{1}{6} \qquad (i = 1, 2, 3, 4, 5, 6)$.

A die was tossed 120 times and the following frequencies occurred.

$o_1 = 17$ $o_2 = 18$ $o_3 = 24$

$o_2 = 26$ $o_5 = 21$ $o_6 = 14$.

Test the hypothesis that the die is fair.

If the hypothesis is true, the expected frequency of each outcome is given by

$$e_i = 120 \times \tfrac{1}{6} = 20 \quad (i = 1, \ldots, 6)$$

Thus we have

$$\chi_0^2 = \frac{(17-20)^2}{20} + \frac{(18-20)^2}{20} + \frac{(24-20)^2}{20}$$
$$+ \frac{(26-20)^2}{20} + \frac{(21-20)^2}{20} + \frac{(14-20)^2}{20}$$
$$= 5 \cdot 1.$$

From Table 3, Appendix B, we have

$$\chi_{0 \cdot 05, 5}^2 = 11 \cdot 07.$$

As χ_0^2 is less than $11 \cdot 07$, the result is not significant at the 5 per cent level and we can accept the null hypothesis that the die is fair.

We have noted that the distribution of the test statistic is only approximately that of a χ^2 random variable. In order to ensure that the approximation is adequate, sufficient observations should be taken so that $e_i \geqslant 5$ for all i. However with less than five categories it is better to have the expected frequencies somewhat larger. If the expected number in any category is too small, the category should be combined with one or more neighbouring categories. (If there are more than about ten categories, then the approximation is valid provided that less than 20 per cent of the values of e_i are less than five, and provided that none is less than one.)

The χ^2 test can also be used to test goodness-of-fit when the null hypothesis depends on unknown parameters which must be estimated from the data. One degree of freedom is deducted for each parameter estimated from the data. Note that it is preferable to use maximum likelihood estimates. The test statistic is often written

$$\sum \frac{(\text{observed} - \text{expected})^2}{\text{expected}}.$$

Example 8

In Example 12, Chapter 4, we saw how to fit a Poisson distribution to the data of Example 2, Chapter 1. At the time we simply noted that there appeared to be good agreement between observed and expected frequencies. We can now confirm this with a χ^2 goodness-of-fit test. The expected Poisson frequencies for $r = 4$ and $r = 5+$ are less than five, and so are too small to be treated separately. They are therefore combined with the results for $r = 3$ to obtain the following observed and expected frequencies.

r	Observed frequency	Poisson frequency
0	13	12·0
1	13	14·4
2	8	8·7
3+	6	4·9

The test statistic is given by

$$\chi_0^2 = \frac{(13-12\cdot0)^2}{12\cdot0} + \frac{(13-14\cdot4)^2}{14\cdot4} + \frac{(9-8\cdot7)^2}{8\cdot7} + \frac{(6-4\cdot9)^2}{4\cdot9}$$

$$= 0\cdot52.$$

We now have two linear restrictions on the frequencies. The sums of the observed and expected frequencies are both equal to forty. In addition the means of the observed distribution and of the fitted Poisson distribution are both equal to 1·2. Therefore, since there are four cells, the number of degrees of freedom is given by $(4-2) = 2$. From Table 3, Appendix B, we have

$$\chi_{0\cdot05,2}^2 = 5\cdot99.$$

The observed value of χ^2 is much smaller than the critical value so we can accept the null hypothesis that the Poisson distribution gives a good fit.

7.5.1   *Testing independence in a two-way table*

A series of observations can often be classified by two types of characteristics into a two-way table.

Example 9

A company manufactures a washing machine at two different factories. A survey is taken on a batch of machines from each factory and a record is kept of whether or not each machine requires a service call during the first six months.

	No service call	*Service required*
Factory A	80	32
Factory B	63	33

Is there evidence that factory A produces more trouble-free appliances than factory B?

The above is an example of a 2×2 two-way table. The general two-way table will have r rows and c columns. If n observations are taken, let n_{ij}, be the number of observations which fall in the ith row and jth column.

Table 12

$r \times c$ Two-way Table

n_{11}	n_{12}	$\cdots$	n_{1c}	$n_1.$
n_{21}	n_{22}	$\cdots$	n_{2c}	$n_2.$
$\vdots$				
n_{r1}	n_{r2}	$\cdots$	n_{rc}	$n_r.$
$n._1$	$n._2$	$\cdots$	$n._c$	n

Let

$$n_{i.} = \sum_j n_{ij} = \text{number of observations in } i\text{th row,}$$

$$n_{.j} = \sum_i n_{ij} = \text{number of observations in } j\text{th column.}$$

Thus

$$n = \sum_i n_{i.} = \sum_j n_{.j}.$$

Generally speaking we are interested in testing the independence of the two types of classification. For this reason two-way tables are often called *contingency tables* because we may ask if the presence of one characteristic is contingent on the presence of another. Thus, in Example 9, if factory A produces a higher proportion of trouble-free appliances than factory B, then a machine is more likely to be 'no call' if it comes from factory A than if it comes from factory B. In such a case the rows and columns are not independent.

We can formalize the null hypothesis as follows. Let p_{ij} be the probability that an item selected at random will be in the ith row and jth column. Let $p_{i\cdot}$ be the probability that an item will be in the ith row and $p_{\cdot j}$ the probability that an item will be in the jth column. Then the null hypothesis that rows and columns are independent is given by

$$H_0 : p_{ij} = p_{i\cdot}p_{\cdot j} \qquad \text{for all } i, j.$$

It is easy to show that the maximum likelihood estimates of $p_{i\cdot}$ and $p_{\cdot j}$ are given by the intuitive estimates $n_{i\cdot}/n$ and $n_{\cdot j}/n$. Thus if H_0 is true an estimate of the expected frequency in the cell in the ith row and jth column is given by

$$n\hat{p}_{ij} = n\hat{p}_{i\cdot}\hat{p}_{\cdot j} = n\frac{n_{i\cdot}}{n}\frac{n_{\cdot j}}{n}$$

$$= \frac{n_{i\cdot}n_{\cdot j}}{n}.$$

This can be compared with the observed value n_{ij}. The test statistic is given by

$$\chi^2 = \sum_{i=1}^{r}\sum_{j=1}^{c}\left\{\frac{(n_{ij} - n_{i\cdot}n_{\cdot j}/n)^2}{n_{i\cdot}n_{\cdot j}/n}\right\}.$$

The number of degrees of freedom is obtained as follows. As the sum of the observed and expected frequencies are equal, this results in the loss of one degree of freedom. In addition the parameters $p_{1\cdot}, p_{2\cdot}, \ldots, p_{r\cdot}$, $p_{\cdot 1}, p_{\cdot 2}, \ldots, p_{\cdot c}$ are estimated from the data. However from the first condition we must have

$$\sum_i p_{i\cdot} = 1 = \sum_j p_{\cdot j}$$

so only $(r + c - 2)$ independent estimates have to be made. Thus the number of degrees of freedom is given by

$$rc - 1 - (r + c - 2) = (r - 1)(c - 1).$$

Example 9 continued

The row and column sums are given by

$$n_{1.} = 112 \quad n_{2.} = 96 \quad n_{.1} = 143 \quad n_{.2} = 65.$$

The total number of observations is given by $n = 208$.

Let e_{ij} denote the expected number of observations in the ith row and jth column. Thus

$$e_{ij} = \frac{n_{i.}n_{.j}}{n}.$$

We find

$$e_{11} = 77 \quad e_{12} = 35 \quad e_{21} = 66 \quad e_{22} = 30.$$

Thus $\quad \chi_0^2 = \dfrac{(80-77)^2}{77} + \dfrac{(32-35)^2}{35} + \dfrac{(63-66)^2}{66} + \dfrac{(33-30)^2}{30}$

$$= 0{\cdot}82.$$

Since number of degrees of freedom $= 1$,

$$\chi_{0{\cdot}05,1}^2 = 3{\cdot}84.$$

Thus the result is not significant at the 5 per cent level and we have no real evidence that factory A produces more trouble-free appliances than factory B.

We have not, as yet, specified an alternative hypothesis when testing independence in a two-way table. If the value of χ^2 for a two-way table is found to be significantly large then the null hypothesis must be rejected. Occasionally we will indeed have a specific alternative hypothesis in mind; but more generally a common procedure is simply to look at the data and see where large discrepancies between observed and expected frequencies occur. This may suggest a suitable hypothesis to describe the data.

7.5.2 *Some remarks on the χ^2 test*

The above comments on the alternative hypothesis also apply in other situations where it is not clear what H_1 is. Thus in Example 8, if we had found that the Poisson distribution did not give a good fit, then we would have had to consider other models for the data, bearing in mind the observed discrepancies from the Poisson model.

In all applications of the χ^2 test the number of degrees of freedom is given by $(k - m)$, where there are k cells and m linear constraints between the observed and expected frequencies. It is often useful to combine the results of successive experiments made at different times on the same problem. This is done by adding the observed values of χ^2 and also adding the number of degrees of freedom from each individual experiment. The total value of χ^2 is then tested with the total number of degrees of freedom. This may give a significant value even though some of the individual values are not significant.

7.6 The F-test

This significance test is widely used for comparing different estimates of variance, particularly in the analysis of variance which is described in Chapter 10.

Suppose that we have a normal distribution with variance σ^2. Two random samples, sizes n_1 and n_2, are drawn from this population and the two sample variances s_1^2 and s_2^2 are calculated in the usual way. As s_1^2 and s_2^2 are both estimates of the same quantity σ^2, we expect the ratio s_1^2/s_2^2 to be 'close' to unity, provided that the samples are reasonably large. If we take repeated pairs of samples, size n_1 and n_2, it can be shown that the ratio $F = s_1^2/s_2^2$ will follow a distribution called the F-distribution.

A random variable which follows the F-distribution can be obtained from two independent χ^2 random variables (see Appendix A). The distribution depends on two parameters, v_1 and v_2, which are the number of degrees of freedom of s_1^2 and s_2^2 respectively. Note that the estimates s_1^2 and s_2^2 must be independent. In the above situation we have $v_1 = n_1 - 1$ and $v_2 = n_2 - 1$.

An F-test is carried out in the following way. Let s_1^2 and s_2^2 be independent estimates of σ_1^2 and σ_2^2 respectively, and assume that the observations in the two samples are normally distributed. Then we are often interested in testing the hypothesis that s_1^2 and s_2^2 are both estimates of the same variance σ^2. In other words we want to test the null hypothesis $H_0 : \sigma_1^2 = \sigma_2^2 = \sigma^2$. For an alternative hypothesis of the form $H_1 : \sigma_1^2 > \sigma_2^2$, a one-tailed test would be appropriate. If the ratio s_1^2/s_2^2 is much greater than one then we will be inclined to reject H_0. In order to see if the observed ratio s_1^2/s_2^2 is significantly large, it is compared with the upper percentage points of the F-distribution which are given in Table 4, Appendix B. The point F_{α, v_1, v_2} is the point on the F-distribution, with v_1 and v_2 d.f., such that a proportion α of the distribution lies above it. If, for example, we find $s_1^2/s_2^2 > F_{0.05, v_1, v_2}$, where s_1^2, s_2^2 are based on v_1, v_2 d.f. respectively, then the result is significant at the 5 per cent level and we have reasonable evidence that H_0 is untrue.

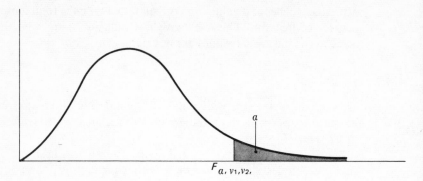

Figure 39   An F-distribution

Occasionally, as in Example 10, the alternative hypothesis is of the form $H_1 : \sigma_1^2 \neq \sigma_2^2$, in which case a two-tailed test is appropriate. In this case the test statistic is chosen so that the larger sample variance is in the numerator, as this enables us to compare it with the upper percentage points of the F-distribution. If this is not done and the test statistic is less than one, then it must be compared with the lower percentage points of the F-distribution, which can be found using the relationship

$$F_{1-\frac{1}{2}\alpha, v_1, v_2} = 1/F_{\frac{1}{2}\alpha, v_2, v_1}.$$

Example 10

In Example 5 we assumed that the two sample variances were estimates of the same population variance. We can now verify this statement with an F-test.

Let s_1^2 be an estimate of σ_1^2, and let s_2^2 be an estimate of σ_2^2. Then the two hypotheses are given by

$$H_0 : \sigma_1^2 = \sigma_2^2,$$

$$H_1 : \sigma_1^2 \neq \sigma_2^2.$$

A two-tailed test is required. We must assume that both populations are normally distributed. From the data we find $s_1^2 = 0.0862$ and $s_2^2 = 0.0672$.

The test statistic is given by

$$F_0 = \frac{s_1^2}{s_2^2} \quad (\text{where } s_1^2 > s_2^2)$$

$$= 1.28.$$

(If we had found $s_1^2 < s_2^2$ the test statistic would have been $F_0 = s_2^2/s_1^2$.)

The number of degrees of freedom of both s_1^2 and s_2^2 is nine. From Table 4, Appendix B, we find $F_{0.025,9,9} = 4.03$, so if s_1^2/s_2^2 were greater than 4.03, the result would be significant at the 5 per cent level. (As we are running a two-tailed test, the level of significance is 2×0.025.) In fact, the observed test statistic is much smaller than 4.03 and we conclude that it is reasonable to accept the null hypothesis.

7.7 Distribution-free or non-parametric tests

In order to apply the z-test or t-test shown in section 7.2, it is necessary to assume that the observations are approximately normally distributed. Occasionally it will not be possible to make this assumption; for example, when the distribution of observations is clearly skewed. Thus a group of tests have been devised in which no assumptions are made about the distribution of the observations. For this reason the tests are called *distribution-free*. Since distributions are compared without the use of parameters, the tests are sometimes called *non-parametric*, though this term can be misleading. The simplest example of a distribution free test is the *sign test* which will be applied to the data of Example 6.

Example 11

The ten differences between the results of method A and method B are

0·1   0·3   0·0   −0·2   0·2   0·3   0·0   0·1   0·1   0·4

Thus seven of the differences have a positive sign, one has a negative sign and two are zero.

Applying the null hypothesis that the two methods give similar results, the differences are equally likely to have a positive or negative sign. If we disregard the two zero differences, the number of positive differences in the remaining eight should follow a binomial distribution with $n = 8$ and $p = \frac{1}{2}$. The expected number of positive differences is given by $np = 4$, but in actual fact there are seven. The level of significance of this result is the probability of observing a result which is as extreme or more extreme. We would be equally suspicious of seven negative differences and even more suspicious of eight positive or eight negative differences. Thus the level of significance is given by

$P(0, 1, 7$ or 8 positive differences)

$$= {}^8C_0(\tfrac{1}{2})^8 + {}^8C_1(\tfrac{1}{2})^8 + {}^8C_7(\tfrac{1}{2})^8 + {}^8C_8(\tfrac{1}{2})^8$$

$$= 0{\cdot}07.$$

Thus the result is not significant at the 5 per cent level.

The result in Example 11 contrasts with the result of Example 6, when a significant result was obtained by using a t-test. Clearly, the conclusions obtained from different significance tests need not always be the same. A general method of comparing significance tests is given in section 7.8. However it is clear that the sign test is not very discriminatory since it takes no account of the magnitude of the differences. If a distribution-free test is required, it is better to use the Wilcoxon signed rank test, which is equivalent to the Mann–Whitney U-test. A description of this test can be found in Wine (1964) or Bennett and Franklin (1954).

In practice, although the true distribution of the observations is seldom known, a preliminary examination of the data is often sufficient to make the assumption that the observations are approximately normally distributed. Moreover it can be shown that small departures from normality do not seriously affect the tests based on the normal assumption which have been described in this chapter. For this reason, these tests are often called *robust*. The tests based on the normal assumption are used much more frequently in the applied sciences than distribution-free tests.

7.8 Power and other considerations

After carrying out a significance test, we have some evidence on which we have to decide whether or not to reject the null hypothesis. Generally speaking H_0 is rejected if the observed value of the test statistic is larger (or smaller) than a particular critical value. This critical value should be chosen before the observations are taken. It is often chosen so that there is at most a 5 per cent chance of rejecting H_0 when it is actually true.

The student is often dismayed by the fact that however the critical value is chosen, it is still possible to make a mistake in two different ways. Firstly it is possible to get a significant result when the null hypothesis is true. This is called an *error of type I*. Secondly it is possible to get a non-significant result when the null hypothesis is false. This is called an *error of type II*.

	H_0 *is true*	H_0 *is false*
Accept H_0	Correct decision	Type II error
Reject H_0	Type I error	Correct decision

Example 12

In Example 1 we showed that in order to test

$$H_0: \mu = 1250$$

against

$$H_1: \mu > 1250$$

it was necessary to consider the test statistic

$$z = \frac{\bar{x} - 1250}{150/\sqrt{25}}.$$

If H_0 is true, this is a standard normal variable and probability $(z > 1.64) = 0.05$.

Thus H_0 is rejected at the 5 per cent level if the observed value of z is greater than 1.64. Therefore 1.64 is the critical value of z. This is equivalent to rejecting H_0 if $\bar{x} > 1250 + (1.64 \times 150)/\sqrt{25} = 1299.2$. So the corresponding critical value of $\bar{x}$ is 1299.2. Here we have chosen the critical value in such a way that the probability of an error of type I is 5 per cent (see Figure 40).

In general the probability of getting a significant result when H_0 is true is denoted by α. The above choice of $\alpha = 0.05$ was quite arbitrary. If changeover costs are high, so that the experimenter requires strong evidence of an improvement, then it would be better to choose $\alpha = 0.01$. If H_0 is true we have

$$\text{probability}\left(\frac{\bar{x} - 1250}{150/\sqrt{25}} < 2.33\right) = 0.99$$

so that the critical value of z would then be 2.33.

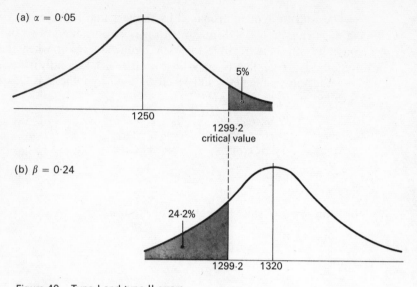

(a) $\alpha = 0.05$

5%

1250

1299·2
critical value

(b) $\beta = 0.24$

24·2%

1299·2   1320

Figure 40   Type I and type II errors
(a) H_0 true: sampling distribution of $\bar{x}$ is $N(1250, 30^2)$
(b) H_1 true: sampling distribution of $\bar{x}$ is $N(1320, 30^2)$

If we have a specific alternative hypothesis and a given value of α, we can also find the probability of an error of type II. This is often denoted by β. For example, let us suppose that the new process really is better and that $\mu = 1320$. Thus $\bar{x}$ would be normally distributed with mean 1320 and standard deviation $150/\sqrt{25}$. But if H_0 is tested with $\alpha = 0.05$, we have seen that the critical value of $\bar{x}$ is 1299·2. If the observed value of $\bar{x}$ exceeds the critical value, then H_0 will be correctly rejected. But a type II error will occur if the observed value of $\bar{x}$ is less than the critical value, so that H_0 is accepted when it is false. If the above alternative hypothesis is true we have

$$\text{probability } (\bar{x} < 1299\cdot2) = \text{probability} \left[\frac{\bar{x} - 1320}{30} < \frac{1299\cdot2 - 1320}{30} \right]$$

$$= \text{probability } (z < -0\cdot695)$$

$$= 0\cdot24.$$

and this is the probability of an error of type II.

We can also find the probability of an error of type II for any other value of α. For example if α is chosen to be 0·01, the critical value of $\bar{x}$

is $1250 + 2.33 \times 30 = 1319.9$. If the above alternative hypothesis is actually true, the chance of getting a non-significant result is given by

$$\text{probability } (\bar{x} < 1319.9) = \text{probability } \left[\frac{\bar{x} - 1320}{30} < \frac{1319.9 - 1320}{30} \right]$$

$$= \text{probability } (z < 0.0)$$

$$= 0.5.$$

Thus, with the lower level of significance, there is a much higher chance of making an error of type II.

From Example 12 it can be seen that the two types of errors are dependent on one another. For example, if the critical value is increased in order to reduce the probability of a type I error, then the probability of a type II error will increase. A 'good' test of significance is one which minimizes the probabilities of these two types of error in some way. For a given value of α we want to choose the significance test which minimizes β.

An alternative way of looking at this problem is to look at the *power* of the test. For a specific alternative hypothesis, the power of a significance test is obtained by calculating the probability that H_0 is rejected when it is false (that is, the correct decision is made). Then

$$\text{power} = 1 - \text{probability (error of type II)}$$

$$= 1 - \beta.$$

For a given value of α, we would like to choose the significance test which has the maximum power.

In general H_1 is often non-specific. For example, if we want to test

$$H_0 : \mu = \mu_0$$

against

$$H_1 : \mu > \mu_0,$$

it is necessary to construct a *power curve* by calculating the probability of rejecting H_0 for different values of μ, and plotting this against μ. The critical value is chosen so that this probability is equal to α when $\mu = \mu_0$.

Suppose we wanted to compare the t-test with some other significance test, such as the sign test. This can be done by constructing the power curves for both tests after choosing the critical values in each test so that each has the same value for α. If the assumptions for the t-test are valid (that is, the sample is randomly selected from a normal population), it can be shown that the power curve of the t-test is always above the power curve of any other test for $\mu > \mu_0$. In other words the t-test is then the most powerful test.

We will not attempt to prove this result, nor will we consider the power of any of the other tests considered in this chapter. In fact it can be shown that all the tests described in this chapter which are based on the normal distribution are most powerful under the stated assumptions. However the χ^2 goodness-of-fit test can have rather poor power properties against certain types of alternative. Hald (1952), Bennett and Franklin (1954), Davies (1958) and Wine (1964) may be consulted for a more detailed discussion of power.

By now the reader should be well aware of the fact that one cannot be certain of making the correct decision as a result of a significance test. It is possible to make an error of type I or of type II. The importance of statistics is that it enables the experimenter to come to a decision in an objective way when faced with experimental uncertainty. However it is always a good idea to give a full statement of the results of an analysis rather than simply to say that a result is significant or non-significant.

We will conclude this chapter by commenting on a topic which was mentioned briefly earlier. So far we have considered situations in which the sample size has been chosen arbitrarily. However this will sometimes mean that the risk associated with a decision is unacceptably large. Thus it is a good idea to choose the sample size in a scientific way.

The following is a typical situation. A random sample is taken from $N(\mu, \sigma^2)$, where σ^2 is known but μ is unknown. It is required to test $H_0 : \mu = \mu_0$ against $H_1 : \mu > \mu_0$. If the sample is of size n the test statistic is given by $z = (\bar{x} - \mu_0)/(\sigma/\sqrt{n})$. The critical value of this statistic can be chosen so that the probability of an error of type I is equal to α. Suppose we also want to ensure that the probability of an error of type II is less than β, if μ is actually equal to $\mu_1(\mu_1 > \mu_0)$. Then it can be shown that the size of the sample must be at least

$$n = \frac{\sigma^2(z_\alpha + z_\beta)^2}{(\mu_1 - \mu_0)^2}.$$

A full discussion of problems of this type is given in reference 1.

A scientific choice of the sample size is also desirable when H_0 is not precise or when the choice of a precise H_0 conflicts with the idea that the burden of proof should be on H_1. For example suppose that a chemist wishes to decide if a certain method of analysis gives unbiased measurements by testing it on several standard solutions of known concentration. There are two rival theories in this situation namely:
(a) The method gives unbiased results.
(b) The method gives biased results.
Here the burden of proof is on hypothesis (a) suggesting that hypothesis (b) should be chosen as H_0. But more statisticians would choose hypothesis (b) as H_0 because it is precise. This doubtful procedure can be remedied by choosing the sample size large enough to give a guaranteed power for a specific bias.

Exercises

1. Test the hypothesis that the random sample

12·1   12·3   11·8   11·9   12·8   12·4

came from a normal population with mean 12·0. The standard deviation of the measurements is known to be 0·4.

Also construct a 95 per cent confidence interval for the true mean, μ.
2. Repeat question one without assuming that the standard deviation is known to be 0·4. In other words estimate the population variance from the sample measurements and use a t-test.
3. A manufacturer claims that the percentage of phosphorus in a fertilizer is at least 3 per cent. Ten small samples are taken from a batch and the percentage of phosphorus in each is measured. The ten measurements have a sample mean of 2·5 per cent and a sample standard deviation of 0·5 per cent. Is this sample mean significantly below the claimed value? State the null hypothesis and the alternative hypothesis and say if a one- or two-tailed test is required.
4. The strength of paper used by a certain company is approximately normally distributed with mean 30 p.s.i. and standard deviation 3 p.s.i. The company decides to test a new source of paper made by a different manufacturer. If this paper is significantly stronger, then the company will switch its trade to this new manufacturer. A batch of paper is

obtained from this manufacturer and a series of measurements are to be made on the strength of different pieces of paper from the batch. Assuming that the standard deviation of these measurements is also 3 p.s.i., how large a sample size should be chosen so as to be 95 per cent certain of detecting a mean increase in strength of 2 p.s.i. with a one-tailed test at the 5 per cent level?

5. For a certain chemical product it is thought that the true percentage of phosphorus is 3 per cent. Ten analyses give $\bar{x} = 3.3$ per cent and $s = 0.2$ per cent. Is the sample mean significantly different from 3 per cent? (This question differs from question 3 because we are interested in departures from 3 per cent in either direction.)

6. One sample of fifteen observations has $\bar{x}_1 = 82$ and $s_1 = 5$. A second sample of ten observations taken by a different scientist has $\bar{x}_2 = 88$ and $s_2 = 7$. Is there a significant difference between the two sample means at the (a) 0.05 and (b) 0.01 level of significance? (You may assume that the two populations have equal variances.)

7. Test the hypothesis that the following set of 200 numbers are 'random digits', that is, each number is equally likely to be 0, 1, 2, ..., 9.

r	0	1	2	3	4	5	6	7	8	9
Frequency	22	16	15	18	16	25	23	17	24	24

8. The following figures show the number of accidents to 647 women in a period of five weeks while working on the manufacture of shells. (Source: M. Greenwood and G. U. Yule, *Journal of the Royal Statistical Society*, 1920.)

Number of accidents	0	1	2	3	4	5	6+
Frequency	447	132	42	21	3	2	0

Find the Poisson distribution with the same mean. Test the hypothesis that the Poisson distribution gives a good fit to the data.

9. In order to test the effectiveness of a new drug in treating a particular disease, seventy patients suffering from the disease were randomly divided into two groups. The first group was treated with the drug and the second group was treated in the standard way. The results were as follows.

	Recover	Die
Drug	20	15
No drug	13	22

Test the hypothesis that the drug has no effect.

References

BENNETT, C. A., and FRANKLIN, N. L. (1954), *Statistical Analysis in Chemistry and the Chemical Industry*, Wiley.

DAVIES, O. L. (ed.) (1958), *Statistical Methods in Research and Production*, Oliver & Boyd, 3rd edn.

HALD, A. (1952), *Statistical Theory with Engineering Applications*, Wiley.

WINE, R. L. (1964), *Statistics for Scientists and Engineers*, Prentice-Hall.

165 Exercises

Chapter 8
Regression and correlation

In previous chapters we have been mainly concerned with the behaviour of one variable without reference to the behaviour of any other variable. In this chapter we consider the situation in which simultaneous measurements are taken on two (or more) variables.

8.1 Scatter diagram

Let us suppose that n pairs of measurements, (x_1, y_1), (x_2, y_2),..., (x_n, y_n), are made on two variables x and y. The first step in the investigation is to plot the data on a scatter diagram in order to get a rough idea of the relationship (if any) between x and y.

Example 1

An experiment was set up to investigate the variation of the specific heat of a certain chemical with temperature. Two measurements of the specific heat were taken at each of a series of temperatures. The following results were obtained.

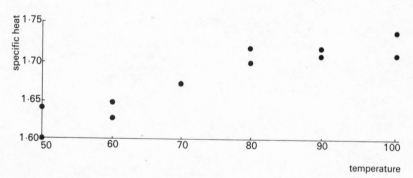

Figure 41   Scatter diagram of data of Example 1

Temperature °C	50	60	70	80	90	100
Specific heat	1·60	1·63	1·67	1·70	1·71	1·71
	1·64	1·65	1·67	1·72	1·72	1·74

Plot the results on a scatter diagram.

8.2 Curve fitting

It is often possible to see, by looking at the scatter diagram, that a smooth curve can be fitted to the data. In particular if a straight line can be fitted to the data then we say that a *linear* relationship exists between the two variables. Otherwise the relationship is *non-linear*.

Situations sometimes occur, particularly in physics and chemistry, in which there is an exact functional relationship between the two variables and in addition the measurement error is very small. In such a case it will usually be sufficiently accurate to draw a smooth curve through the observed points by eye. Here there is very little experimental uncertainty and no statistical analysis is really required.

However most situations are not so clear cut as this, and then a more systematic method is required to find the relationship between the two variables. In the first part of this chapter we will discuss the situation where the values of one of the variables are determined by the experimenter. This is called the controlled, independent or regressor variable. The resulting value of the second variable depends on the selected value of the controlled variable. Therefore the second variable is called the dependent or response variable. However the problem is usually complicated by the fact that the dependent variable is subject to a certain amount of experimental variation or scatter.

Thus, in Example 1, the temperature is the controlled variable and the specific heat is the dependent variable. At a fixed temperature, the two observations on the specific heat vary somewhat. Nevertheless it can be seen that the average value of the specific heat increases with the temperature.

The problem now is to fit a line or curve to the data in order to predict the mean value of the dependent variable for a given value of the controlled variable. If the dependent variable is denoted by y and the controlled variable by x, this curve is called the *regression* curve, or line, of y on x.

We will begin by considering the problem of fitting a straight line to n pairs of measurements, $(x_1, y_1), \ldots, (x_n, y_n)$, where the y_i are subject to scatter but the x_i are not. A straight line can be represented by the equation

$$y = a_0 + a_1 x.$$

Our task is to find estimates of a_0 and a_1 such that the line gives a good fit to the data. One way of doing this is by the '*method of least squares*'. At any point x_i the corresponding point on the line is given by $a_0 + a_1 x_i$, so the difference between the observed value of y and the predicted value is given by

$$e_i = y_i - (a_0 + a_1 x_i).$$

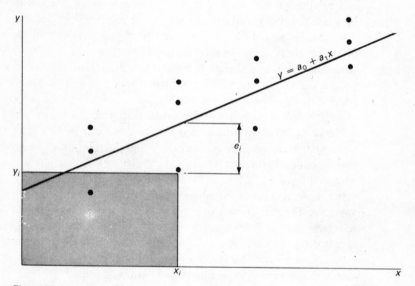

Figure 42

The least squares estimates of a_0 and a_1 are obtained by choosing the values which minimize the sum of squares of these deviations.

The sum of the squared deviations is given by

$$S = \sum_{i=1}^{n} e_i^2$$

$$= \sum_{i=1}^{n} [y_i - (a_0 + a_1 x_i)]^2.$$

This quantity is a function of the unknown parameters a_0 and a_1. It can be minimized by calculating $\partial S/\partial a_0$ and $\partial S/\partial a_1$, setting both these partial derivatives equal to zero, and solving the two simultaneous equations to obtain the least squares estimates, $\hat{a}_0$ and $\hat{a}_1$, of a_0 or a_1.

We have

$$\frac{\partial S}{\partial a_0} = \sum 2(y_i - a_0 - a_1 x_i)(-1),$$

$$\frac{\partial S}{\partial a_1} = \sum 2(y_i - a_0 - a_1 x_i)(-x_i).$$

When these are both zero we have

$$\sum 2(y_i - \hat{a}_0 - \hat{a}_1 x_i)(-1) = 0,$$

$$\sum 2(y_i - \hat{a}_0 - \hat{a}_1 x_i)(-x_i) = 0.$$

These can be rearranged to give

$$n\hat{a}_0 + \hat{a}_1 \sum x_i = \sum y_i,$$
$$\hat{a}_0 \sum x_i + \hat{a}_1 \sum x_i^2 = \sum x_i y_i.$$

8.1

These two simultaneous equations in $\hat{a}_0$ and $\hat{a}_1$ are often called the normal equations. They can be solved to give

$$\hat{a}_0 = \bar{y} - \hat{a}_1 \bar{x},$$

$$\hat{a}_1 = \frac{\sum (x_i - \bar{x})(y_i - \bar{y})}{\sum (x_i - \bar{x})^2}.$$

It is easy to see that the least squares regression line passes through the centroid of the data, $(\bar{x}, \bar{y})$, with slope $\hat{a}_1$. Thus it is often convenient to write the equation in the equivalent form

$$y - \bar{y} = \hat{a}_1(x - \bar{x}).$$

If an electric calculating machine is available, it is easier to calculate $\hat{a}_1$ using the equivalent formula

$$\hat{a}_1 = \frac{n \sum x_i y_i - \sum y_i \sum x_i}{n \sum x_i^2 - (\sum x_i)^2}.$$

Each of the required quantities, namely $\sum x_i$, $\sum y_i$, $\sum x_i^2$ and $\sum x_i y_i$, can be easily computed from the data.

The formula for $\hat{a}_1$ is also equivalent to

$$\hat{a}_1 = \frac{\sum x_i y_i - n\bar{x}\bar{y}}{\sum x_i^2 - n\bar{x}^2}, \qquad\qquad \textbf{8.2}$$

which can be used if $\bar{x}$ and $\bar{y}$ have already been calculated.

After the least squares regression line has been calculated, it is possible to predict values of the dependent variable. At a particular value, x_0, of the controlled variable, the point estimate of y is given by $\hat{a}_0 + \hat{a}_1 x_0$.

Example 2

Fit a straight line to the data of Example 1 and estimate the specific heat when the temperature is 75°C.

By inspection we have $\bar{x} = 75 =$ average temperature.

By calculation we have $\sum_{i=1}^{12} y_i = 20\cdot16$ giving $\bar{y} = 1\cdot68$.

We also have $\sum y^2 = 33\cdot8894$, $\sum xy = 1519\cdot9$ and $\sum x^2 = 71,000$. Note particularly in the last summation that each value of x must be considered twice as there are two measurements on the specific heat at each temperature.

From equation **8.2** we find

$$\hat{a}_1 = 0\cdot00226.$$

Hence $\hat{a}_0 = 1\cdot51.$

Thus the estimated regression line of y on x is given by

$$y = 1\cdot51 + 0\cdot00226x.$$

When $x = 75$ the prediction of the specific heat is given by

$$1\cdot51 + 0\cdot00226 \times 75 = 1\cdot68.$$

In many cases an inspection of the scatter diagram is sufficient to see that the regression equation is not a straight line. One way of using linear theory in the non-linear case is to transform the variables in such a way that a linear relationship results. For example, if two variables are related by the formula

$$y = a_0 x^c,$$

then we have

$$\log y = \log a_0 + c \log x,$$

so that if $\log y$ is plotted against $\log x$, the points will lie on a straight line. One advantage of such a transformation is that it is easier to fit a straight line than any other type of curve. A full account of such transformations is given by Hald (1952) section 18.7.

In general it will more often be necessary to try and fit a non-linear curve to the data. Before describing how this is done, we will discuss the problem of regression in more general terms.

8.3 Regression

We have seen that if several measurements are made on the dependent variable, y, at the same value of the controlled variable, x, then the results will form a distribution. The curve which joins the mean values of these distributions is called the *regression curve* of y on x, and an example is given in Figure 43. The problem of finding the most suitable form of equation to predict one variable from the values of one, or more, other variables is called the problem of *regression*.

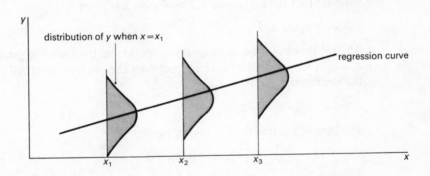

Figure 43 Regression curve. The locus of the mean values of the y-distributions

In order to estimate the regression curve of y on x we must first specify the *functional form* of the curve. Some examples are

$y = a_0 + a_1 x$ linear regression,

$y = a_0 + a_1 x + a_2 x^2$ quadratic regression,

$y = a_0 e^x.$

The functional form may be selected from theoretical considerations. For example, the experiment may have been designed to verify a particular relationship between the variables. Alternatively the functional form may be selected after inspecting the scatter diagram, as it would be pointless for example to try and fit a straight line to some data if the relationship was clearly non-linear. In practice the experimenter may find it necessary to fit several different types of curve to the data and to choose the one which gives the best fit (see section 8.7).

After the functional form has been selected, the next problem is to estimate the unknown parameters of the curve. If, for example, a quadratic relationship is thought to be appropriate, the regression curve is given by

$$y = a_0 + a_1 x + a_2 x^2$$

and then the quantities a_0, a_1 and a_2 must be estimated from the data.

A general method of estimating the parameters of a regression curve is by the method of least squares. We have already described how this is done when the regression curve is a straight line. A similar technique can be adopted with other regression curves. For example, let us suppose that the regression curve of y on x is given by

$$y = a_0 + a_1 x + a_2 x^2.$$

At any point x_i, the corresponding point on the curve is given by $a_0 + a_1 x_i + a_2 x_i^2$, so the difference between the observed value of y and the predicted value is

$$e_i = y_i - (a_0 + a_1 x_i + a_2 x_i^2).$$

The sum of squared deviations is given by

$$S = \sum e_i^2 = \sum (y_i - a_0 - a_1 x_i - a_2 x_i)^2.$$

This quantity can be minimized by calculating $\partial S/\partial a_0$, $\partial S/\partial a_1$ and $\partial S/\partial a_2$, and setting the partial derivatives equal to zero. Then the least squares estimates can be obtained by solving the resulting simultaneous equations which are

$$\left.\begin{aligned}
n\hat{a}_0 + \hat{a}_1 \sum x_i + \hat{a}_2 \sum x_i^2 &= \sum y_i, \\
\hat{a}_0 \sum x_i + \hat{a}_1 \sum x_i^2 + \hat{a}_2 \sum x_i^3 &= \sum y_i x_i, \\
\hat{a}_0 \sum x_i^2 + \hat{a}_1 \sum x_i^3 + \hat{a}_2 \sum x_i^4 &= \sum y_i x_i^2.
\end{aligned}\right\} \qquad \textbf{8.3}$$

These are the normal equations for quadratic regression. They can be solved to give the least squares estimates $\hat{a}_0$, $\hat{a}_1$ and $\hat{a}_2$.

The next problem that arises is to find the conditions under which the least squares estimates are 'good' estimates of the unknown parameters of the regression equation. In the case of linear regression it can be shown that the least squares estimates are maximum likelihood estimates if the following conditions apply:

(1) For a fixed value of the controlled variable, x_0 say, the dependent variable follows a normal distribution with mean $a_0 + a_1 x_0$.
(2) The conditional variance of the distribution of y for a fixed value of x is a constant, usually denoted by $\sigma^2_{x|y}$. Thus the conditional variance of the dependent variable does not depend on the value of x.

A preliminary examination of the data should be made to see if it is reasonable to assume that these conditions do apply. Together these assumptions constitute what is called the *linear regression model*. This model states that for a fixed value of x, say x_0, the dependent variable, y, is a random variable such that

$$E(y|x) = a_0 + a_1 x.$$

Thus the line

$$y = a_0 + a_1 x$$

joins the mean values of the y-distributions and is the true regression line. In order to distinguish the true regression line from the estimated regression line we will denote the latter by

$$\hat{y} = \hat{a}_0 + \hat{a}_1 x.$$

The assumptions made in the above model are also necessary to establish confidence intervals for the true values of a_0 and a_1, and for the mean values of the y-distributions. This will be described in the next section.

In general the properties of normality and constant variance are often important assumptions in many other types of model. For example, in quadratic regression, suppose that the dependent variable follows a normal distribution with mean $a_0 + a_1 x + a_2 x^2$, where x is the value of the controlled variable. If the variance of the y-distributions is a constant $\sigma^2_{y|x}$ which does not depend on the value of x, then the least squares estimates of a_0, a_1 and a_2 are in fact the maximum likelihood estimates.

Note that if the variance of the y-distributions does vary with x, then the least squares procedure must be modified by giving more weight to those observations which have the smaller variance. This problem is considered by Hald (1952) section 18.6. A somewhat similar problem is also considered in this book in section 9.5.

8.4 Confidence intervals and significance tests in linear regression

The next two sections will be concerned with completing the discussion of linear regression. Once we have made the assumptions of normality and constant variance, as described in the previous section, we are in a position to obtain confidence intervals for a_0, a_1 and $a_0 + a_1 x$. The results will be stated without formal proofs.

Given n pairs of observations, $(x_1, y_1), \ldots, (x_n, y_n)$, the least squares estimates of a_0 and a_1 can be obtained in the manner described in section 8.2. These estimates can be expected to vary somewhat from sample to sample. However it can be shown that both estimates are unbiased and hence that $\hat{a}_0 + \hat{a}_1 x$ is an unbiased estimate of $a_0 + a_1 x$. In addition it can be shown that

$$\text{variance}(\hat{a}_0) = \frac{\sigma_{y|x}^2}{n}\left[1 + \frac{n\bar{x}^2}{\sum(x_i - \bar{x})^2}\right],$$

that $\quad \text{variance}(\hat{a}_1) = \dfrac{\sigma_{y|x}^2}{\sum(x_i - \bar{x})^2}$

and that both $\hat{a}_0$ and $\hat{a}_1$ are normally distributed, as each is a linear combination of the observed values of y which are themselves normally distributed. Thus in order to obtain confidence intervals for a_0, a_1 and $a_0 + a_1 x$ we must first obtain an estimate of the residual variance, $\sigma_{y|x}^2$.

The sum of squared deviations of the observed points from the estimated regression line is given by

$$\sum(y_i - \hat{a}_0 - \hat{a}_1 x_i)^2.$$

This quantity is often called the residual sum of squares. It can be shown that an unbiased estimate of $\sigma_{y|x}^2$ can be obtained by dividing this residual sum of squares by $n - 2$

$$s_{y|x}^2 = \frac{\sum(y_i - \hat{a}_0 - \hat{a}_1 x_i)^2}{n - 2}.$$

The denominator, $n - 2$, shows that two degrees of freedom have been lost. This is because the two quantities $\hat{a}_0$ and $\hat{a}_1$ were estimated from the data, so there are two linear restrictions on the values of $y_i - \hat{a}_0 - \hat{a}_1 x_i$.

A more convenient form for computation is given by

$$s_{y|x}^2 = \frac{\sum y_i^2 - \hat{a}_0 \sum y_i - \hat{a}_1 \sum x_i y_i}{n-2}.$$

Then it can be shown that the $100(1-\alpha)$ per cent confidence interval for a_1 is given by

$$\hat{a}_1 \pm t_{\frac{1}{2}\alpha, n-2} \times \frac{s_{y|x}}{\sqrt{[\sum(x_i - \bar{x})^2]}},$$

for a_0 by

$$\hat{a}_0 \pm t_{\frac{1}{2}\alpha, n-2} \times s_{y|x} \sqrt{\left[\frac{1}{n} + \frac{\bar{x}^2}{\sum(x_i - \bar{x})^2}\right]}$$

and for $a_0 + a_1 x_0$ by

$$\hat{a}_0 + \hat{a}_1 x_0 \pm t_{\frac{1}{2}\alpha, n-2} \times s_{y|x} \sqrt{\left[\frac{1}{n} + \frac{(x_0 - \bar{x})^2}{\sum(x_i - \bar{x})^2}\right]}.$$

The confidence interval for $a_0 + a_1 x$ is illustrated in Figure 44. Note that it is shortest when $x_0 = \bar{x}$.

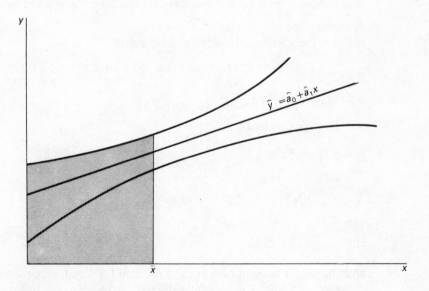

Figure 44   Confidence intervals for $(a_0 + a_1 x)$

175   Confidence intervals and significance tests in linear regression

Most engineers believe that the best test procedure for establishing a relation between two variables is to run equally spaced observations. This approach is in fact quite correct if there is no prior information about the relationship, particularly if it is intended to use the method of orthogonal polynomials to analyse the data, as shown in section 8.7. However if the experimenter already has convincing evidence that the relationship is linear, then the above results indicate that it is better to take more observations at the ends and 'starve' the middle. This will increase the value of $\sum(x_i - \bar{x})^2$ and so decrease the standard error of all the above estimates.

One question which frequently arises is whether or not the slope of the regression line is significantly different from zero. In other words it is desired to test the hypothesis $H_0 : a_1 = 0$ against the alternative hypothesis $H_1 : a_1 \neq 0$. The test statistic is given by

$$\frac{\hat{a}_1 \sqrt{\left[\sum(x_i - \bar{x})^2\right]}}{s_{y|x}}$$

and follows the t-distribution with $n - 2$ degrees of freedom if H_0 is true.

Example 3

In Example 2 the estimated slope of the regression line is given by

$\hat{a}_1 = 0 \cdot 00226.$
Is this value significantly different from zero?

We have $\quad s_{y|x}^2 = \dfrac{33 \cdot 8894 - 1 \cdot 510833 \times 20 \cdot 16 - 0 \cdot 00225556 \times 1519 \cdot 9}{10}$

$\qquad\qquad = 0 \cdot 00034,$

therefore $\quad s_{y|x} = 0 \cdot 018,$

$\sum(x_i - \bar{x})^2 = 3500.$

$$t_0 = \frac{0 \cdot 00226 \times 59 \cdot 1}{0 \cdot 018}$$

$= 7 \cdot 3.$

But $\quad t_{0 \cdot 025, 10} = 2 \cdot 23.$

Thus the result is significant at the 5 per cent level and so the slope of the regression line is significantly different from zero, even though it appears, at first sight, to be very small.

We have seen how to find a confidence interval for the mean value of y, for a given value of x. However we are often more interested in the spread of the observations around this mean value. Thus it would be useful to find an interval in which future values of y will probably lie. It can be shown that there is a probability $1 - \alpha$ that a future observation on y, at the point x_0, will lie between

$$\hat{a}_0 + \hat{a}_1 x_0 \pm t_{\frac{1}{2}\alpha, n-2}\, s_{y|x} \sqrt{\left[1 + \frac{1}{n} + \frac{(x_0 - \bar{x})^2}{\sum\limits_{i=1}^{n}(x_i - \bar{x})^2}\right]}.$$

This interval is often called a *prediction interval*.

8.5   **The coefficient of determination**

Another important consideration is to see how well the estimated regression line fits the data. In order to achieve this we will use the following important relationship. If the least squares line is given by

$$\hat{y} = \hat{a}_0 + \hat{a}_1 x,$$

it can be shown that

$$\sum_{i=1}^{n}(y_i - \bar{y})^2 = \sum_{i=1}^{n}(y_i - \hat{y}_i)^2 + \sum_{i=1}^{n}(\hat{y}_i - \bar{y})^2.$$

The quantity $\sum(y_i - \bar{y})^2$ measures the sum of squared deviations of the observed y-values from $\bar{y}$. This is often called the total corrected sum of squares of y or the '*total variation*' in y. The quantity $\sum(y_i - \hat{y}_i)^2$ is the residual sum of squares and was introduced in section 8.4. It is sometimes called the '*unexplained variation*'. The quantity $\sum(\hat{y}_i - \bar{y})^2$ represents the variation of the points $\hat{y}_i$ on the estimated regression line and is often called the '*explained variation*'. The important point is that the total variation can be partitioned into two components, the explained and unexplained variation.

Total variation = explained variation + unexplained variation.

The ratio of the explained variation to the total variation measures how well the straight line fits the data. This ratio is called the *coefficient of determination* and must lie between nought and one. If it is equal to one, then all the observed points lie exactly on a straight line. If it is equal to nought, then $\hat{y}_i = \bar{y}$ for all i, so the slope of the regression line is zero. Therefore the closer the coefficient is to one, the closer the points lie to an exact straight line. The coefficient can also be calculated in a similar way for a non-linear curve, which is fitted to a set of data by the method of least squares. The total variation is partitioned into the explained variation and the residual sum of squares in a similar way.

The coefficient of determination for linear regression turns out to be the square of a quantity called the correlation coefficient which is considered in section 8.9.

8.6 Multiple and curvilinear regression

The analysis of the linear regression model can be extended in a straightforward way to cover situations in which the dependent variable is affected by several controlled variables, or in which it is affected non-linearly by one controlled variable.

For example, suppose that there are three controlled variables, x_1, x_2, and x_3. A linear regression equation is of the form

$$y = a_0 + a_1 x_1 + a_2 x_2 + a_3 x_3.$$

Given n sets of measurements, $(y_1, x_{11}, x_{21}, x_{31}), \ldots, (y_n, x_{1n}, x_{2n}, x_{3n})$, the least squares estimates of a_0, a_1, a_2 and a_3 can be obtained in a similar way to that previously described. The sum of squared deviations of the observed values of y from the predicted values is given by

$$S = \sum (y_i - a_0 - a_1 x_{1i} - a_2 x_{2i} - a_3 x_{3i})^2.$$

This quantity can be minimized by setting $\dfrac{\partial S}{\partial a_0}$, $\dfrac{\partial S}{\partial a_1}$, $\dfrac{\partial S}{\partial a_2}$ and $\dfrac{\partial S}{\partial a_3}$ equal to zero, to obtain four simultaneous equations in $\hat{a}_0$, $\hat{a}_1$, $\hat{a}_2$ and $\hat{a}_3$.

$$\hat{a}_0 n + \hat{a}_1 \sum x_{1i} + \hat{a}_2 \sum x_{2i} + \hat{a}_3 \sum x_{3i} = \sum y_i,$$

$$\hat{a}_0 \sum x_{1i} + \hat{a}_1 \sum x_{1i}^2 + \hat{a}_2 \sum x_{2i} x_{1i} + \hat{a}_3 \sum x_{3i} x_{1i} = \sum y_i x_{1i},$$

$$\hat{a}_0 \sum x_{2i} + \hat{a}_1 \sum x_{1i} x_{2i} + \hat{a}_2 \sum x_{2i}^2 + \hat{a}_3 \sum x_{3i} x_{2i} = \sum y_i x_{2i},$$

$$\hat{a}_0 \sum x_{3i} + \hat{a}_1 \sum x_{1i} x_{3i} + \hat{a}_2 \sum x_{2i} x_{3i} + \hat{a}_3 \sum x_{3i}^2 = \sum y_i x_{3i}.$$

8.4

These four equations, called the normal equations, can be solved to give the least squares estimates of a_0, a_1, a_2 and a_3. The simplest method of solving equations of this type is by successive elimination (see Hald, 1952). Details of this technique will not be given here as general multiple regression programs are now available with most computers, and the reader is advised to make use of them if three or more controlled variables are present.

To derive confidence intervals for the regression parameters and for the predicted values, it is necessary to make the assumptions of normality and constant variance as in linear regression. The reader is referred to Hald (1952).

Curvilinear regression, which has already been referred to in section 8.3, is tackled in a similar way to multiple regression and can, if desired, be considered as a special case of multiple regression. We will now discuss some special cases.

8.6.1 Polynomial regression

Let us suppose that the dependent variable is a polynomial function of a single controlled variable. For example, in cubic regression, the regression equation is given by

$$y = a_0 + a_1 x + a_2 x^2 + a_3 x^3.$$

This type of regression can be approached in the same way as multiple regression. In the case of cubic regression we can substitute $x_1 = x$, $x_2 = x^2$ and $x_3 = x^3$. The least squares estimates of a_0, a_1, a_2 and a_3 can then be obtained from the normal equations **8.4**. However these normal equations involve terms like $\sum y_i x_i^3$ and are tedious to solve.

If the observations are taken in such a way that there are an equal number of observations on y at a series of equally spaced values of x, then it is easier to use the method of orthogonal polynomials, which is described in section 8.7.

8.6.2 Mixtures

Some regression models involve a combination of multiple and curvilinear regression. Two examples of this are

$$y = a_0 + a_1 x + a_2 x^2 + a_3 z,$$

$$y = a_0 + a_1 x + a_2 z + a_3 xz.$$

In the second equation the term $a_3 xz$ implies an *interaction* between the two controlled variables x and z. Both situations can be analysed in the manner previously described. In the first case we set $x_1 = x$, $x_2 = x^2$ and $x_3 = z$; in the second case we set $x_1 = x$, $x_2 = z$ and $x_3 = xz$.

8.6.3 *Transformations*

Theoretical considerations may lead to a regression model which depends on a transformation of the controlled variables. For example, if the regression equation is of the form

$$y = a_0 + a_1 \log x + a_2 \log z,$$

where x and z are the controlled variables, then estimates of a_0, a_1 and a_2 can be obtained by setting $x_1 = \log x$ and $x_2 = \log z$.

8.7 Orthogonal polynomials

The amount of computation involved in polynomial regression may be substantial, and it increases rapidly with the order of the polynomial. Therefore, if possible, we would like to plan the experiment so that short-cut techniques can be employed. Fortunately if the values of the controlled variable are equally spaced and an equal number of observations are made at each point, then the method of *orthogonal polynomials* can be used. This considerably reduces the amount of computation.

For polynomial regression of degree k the regression equation is usually written in the form

$$y = a_0 + a_1 x + \ldots + a_k x^k.$$

To use the method of orthogonal polynomials this equation is rewritten in the following form:

$$y = a_0'' + a_1'' f_1(x) + \ldots + a_k'' f_k(x),$$

where $f_r(x)$ is a polynomial in x of degree r, and the constants a_i'' depend on the values of a_i.

It can be shown that it is always possible to find a set of polynomials which satisfy the following conditions:

$$\sum_x f_r(x) f_s(x) = 0 \quad (r \neq s),$$

$$\sum_x f_r(x) = 0.$$

These polynomials are called orthogonal polynomials. They depend on the average value, $\bar{x}$, of the controlled variable and on d, the distance between successive values of x.

The method of calculating these polynomials is rather complicated. In practice it is much easier to work with a standardized (coded) controlled variable which is given by

$$z = \frac{x - \bar{x}}{d}.$$

The corresponding regression equation is of the form

$$y = a'_0 + a'_1 f_1(z) + \ldots + a'_k f_k(z).$$

These standardized orthogonal polynomials are tabulated in both Fisher and Yates (1963) and Pearson and Hartley (1966). The standardized controlled variable is symmetric about zero and there is a unit distance between successive values of z.

The next step is to obtain least squares estimates of $a'_0, a'_1, \ldots, a'_k$. First of all we consider the case where one observation is made on y at n different values of x (or z). If the normal equations are derived as in the previous section, it is found that they reduce to the following simple equations

$$n\hat{a}'_0 = \sum y_i$$

$$\hat{a}'_1 \sum f_1(z_i)^2 = \sum f_1(z_i) y_i$$

$$\vdots$$

$$\hat{a}'_k \sum f_k(z_i)^2 = \sum f_k(z_i) y_i.$$

All the other terms are zero because the polynomials are orthogonal; all summations are for $i = 1$ to n. The quantities $\sum f_r(z_i)^2$ are given in the above tables. Thus the only quantities which have to be calculated are $\sum f_r(z_i) y_i$; $r = 1$ to k. This can easily be done as the numerical values of $f_r(z_i)$ are also given in Fisher and Yates (1963) and Pearson and Hartley (1966). The estimates follow immediately.

In general if c observations on y are made for n different values of x then the normal equations are modified by multiplying the terms on the left hand side of the equations by c. Finally the regression equation can be obtained as a polynomial in x by substituting

$$z = \frac{(x - \bar{x})}{d}.$$

Another advantage of orthogonal polynomials is that the estimates of the regression parameters are independent. This fact is particularly useful when the order of the polynomial is not known beforehand. The problem then is to find the lowest order polynomial which fits the data adequately. In this case it is best to adopt a sequential approach, by first fitting a straight line, then fitting a quadratic curve, then a cubic curve, and so on. At each stage it is only necessary to estimate one additional parameter as the earlier estimates do not change.

At each stage an F-test is performed to test the adequacy of the fit. In order to describe this we will again assume that one observation is made at n different values of x. The results are stated without proof. After fitting a straight line, the residual sum of squares is given by

$$R_1 = \sum [y - \hat{a}'_0 - \hat{a}'_1 f_1(z)]^2.$$

After fitting a quadratic curve, this is reduced to

$$R_2 = \sum [y - \hat{a}'_0 - \hat{a}'_1 f_1(z) - \hat{a}'_2 f_2(z)]^2.$$

It can be shown that this reduction is given by

$$R_1 - R_2 = \hat{a}'^2_2 \sum f_2(z)^2.$$

In general, after fitting a polynomial of degree k, the residual sum of squares is given by

$$R_k = \sum [y - \hat{a}'_0 - \hat{a}'_1 f_1(z) \ldots - \hat{a}'_k f_k(z)]^2$$
$$= \sum (y - \bar{y})^2 - \hat{a}'^2_1 \sum f_1(z)^2 - \ldots - \hat{a}'^2_k \sum f_k(z)^2.$$

The residual sum of squares involves $(k+1)$ parameters which have been estimated from the data. In order to estimate the residual variance we would divide this quantity by $n - (k+1)$, which is the number of degrees of freedom corresponding to R_k. We say that R_k is on $n - (k+1)$ degrees of freedom. Now at stage j the residual sum of squares is reduced by $\hat{a}'^2_j \sum f_j(z)^2$, a quantity which is on one degree of freedom. If a polynomial of degree $k-1$ fits the data adequately then $R_{k-1}/(n-k)$ will be an estimate of the residual variance $\sigma^2_{y|x}$. But R_{k-1} can be split up into two components R_k and $\hat{a}'^2_k \sum f_k(z)^2$ so that $R_k/(n-k-1)$ and $\hat{a}'^2_k \sum f_k(z)^2$ will then be independent estimates of $\sigma^2_{y|x}$. Then the ratio of these two quantities will follow an F-distribution with one and $n-k-1$ degrees of freedom.

$$F = \frac{\hat{a}'^2_k \sum f_k(z)^2}{R_k/(n-k-1)}.$$

On the other hand if a polynomial of degree k does fit the data better than a polynomial of degree $(k-1)$, then R_k will be substantially less than R_{k-1}, and the F-ratio may be significantly large.

This sequential process is continued until two non-significant F-ratios in a row are obtained. This is necessary because even-order polynomials may give non-significant results even though odd-order polynomials do give significant results – and vice versa.

The method is illustrated in Example 4.

Example 4

Find the polynomial of the lowest degree which adequately describes the following hypothetical data in which x is the controlled variable.

x	0	1	2	3	4	5	6
y	6·3	5·7	6·3	7·3	9·9	12·5	18·1

The standardized controlled variable is given by

$$z = x - 3.$$

There are $n = 7$ values of the controlled variable. From Pearson and Hartley (1966), the first four orthogonal polynomials are the following:

z	$f_1(z)$	$f_2(z)$	$f_3(z)$	$f_4(z)$
-3	-3	5	-1	3
-2	-2	0	1	-7
-1	-1	-3	1	1
0	0	-4	0	6
$+1$	$+1$	-3	-1	1
$+2$	$+2$	0	-1	-7
$+3$	$+3$	5	1	3
$\sum f(z)^2$	28	84	6	154

Then we find

$$\sum_{i=1}^{7} f_1(z_i)y_i = 52 \cdot 6,$$

$$\sum f_2(z)y = 44 \cdot 2,$$

$$\sum f_3(z)y = 1 \cdot 4,$$

$$\sum f_4(z)y = 5 \cdot 8,$$

$$\sum y = 66 \cdot 1.$$

Thus we have

$$\hat{a}_0' = \bar{y} = 9 \cdot 44,$$

$$\hat{a}_1' = \frac{52 \cdot 6}{28} = 1 \cdot 878,$$

$$\hat{a}_2' = \frac{44 \cdot 2}{84} = 0 \cdot 526,$$

$$\hat{a}_3' = \frac{1\cdot4}{6} = 0\cdot233,$$

$$\hat{a}_4' = \frac{5\cdot8}{154} = 0\cdot037.$$

The next stage is to compute a series of F-ratios to see how many of these parameters are required. At each stage two mean squares are obtained by dividing the appropriate sum of squares by the appropriate number of degrees of freedom. The ratio of the mean squares, the F-ratio, is then compared with $F_{0\cdot05,1,n-k-1}$.

Table 13

Type of variation	Sum of squares	d.f.	Mean square	F-ratio	$F_{0\cdot05}$
Residual from mean	$\sum(y-\bar{y})^2 = 122\cdot86$	6			
Explained by linear	$\hat{a}_1'^2\sum f_1(z)^2 = 98\cdot72$	1	98·72	20·6	6·6
Residual from linear	24·14	5	4·82		
Explained by quadratic	$\hat{a}_2'^2\sum f_2(z)^2 = 23\cdot24$	1	23·24	105·6	7·7
Residual from quadratic	0·90	4	0·22		
Explained by cubic	$\hat{a}_3'^2\sum f_3(z)^2 = 0\cdot32$	1	0·32	1·7	10·1
Residual from cubic	0·58	3	0·19		
Explained by quartic	$\hat{a}_4'^2\sum f_4(z)^2 = 0\cdot21$	1	0·21	1·1	18·5
Residual from quartic	0·37	2	0·19		

Neither the cubic nor the quartic terms give a significant F-ratio. In any case, the residual sum of squares is so small after fitting linear and quadratic terms that it is really unnecessary to try higher order terms in this case. Thus a quadratic polynomial describes the data adequately.

The next stage is to compute the regression equation in terms of the original controlled variable, x. For this we need to know the orthogonal polynomials as functions of z. These are also given in Fisher and Yates (1963) and Pearson and Hartley (1966). We find

$$f_1(z) = \lambda_1 z \quad \text{with} \quad \lambda_1 = 1$$

and

$$f_2(z) = \lambda_2(z^2 - 4) \quad \text{with} \quad \lambda_2 = 1.$$

Thus the estimated regression equation is given by

$$y' = 9\cdot44 + 1\cdot88z + 0\cdot53(z^2 - 4)$$
$$= 9\cdot44 + 1\cdot88(x - 3) + 0\cdot53[(x - 3)^2 - 4]$$
$$= 6\cdot45 - 1\cdot30x + 0\cdot53x^2.$$

8.8 The design of regression experiments

So far, we have said little about how to plan a regression experiment, although this is a very important topic. Here we will only make a few preliminary remarks as the design of experiments is considered in some detail in Chapters 10 and 11. Nevertheless it is important to realise that a little foresight while the data is being collected can lead to a substantial decrease in the amount of computation required. A good design is also necessary to ensure that the conclusions from the experiment are valid.

In polynomial regression, we have already seen that if successive values of the controlled variable are an equal distance apart, then the method of orthogonal polynomials can be used. Even in linear regression a similar restriction on the values of the controlled variable will reduce the amount of arithmetic (see exercise 2). In multiple regression we will see that the best design is one in which observations are made at a rectangular grid of points. This experiment is called a complete factorial experiment. After the values of the controlled variables have been standardized, the cross product terms in the normal equations turn out to be zero and so the equations become much easier to solve (see Example 3, Chapter 11).

The above remarks have been concerned with reducing the amount of arithmetic. Another consideration is to carry out the experiment in such a way that nuisance factors do not affect the results of the experiment. This can usually be achieved by randomizing the order of the experiments. A full discussion of this technique is given in Chapter 10.

8.9 The correlation coefficient

The first part of this chapter has been concerned with the problem of regression. We now turn our attention to a different type of situation in which measurements are made simultaneously on two variables, neither of which can be controlled. In other words they are both random variables.

Some typical data of this type is given in Example 5.

Example 5

The following pairs of (coded) measurements were taken of the temperature and thrust of a rocket engine while it was being run under the same operating conditions. Plot the results on a scatter diagram.

x	y	x	y	x	y
19	1·2	33	2·1	45	2·2
15	1·5	30	2·5	39	2·2
35	1·5	57	3·2	25	1·9
52	3·3	49	2·8	40	1·8
35	2·5	26	1·5	40	2·8

Data of this type occurs frequently in biology and the social sciences. For example measurements are often made on two different characteristics of the same human being, such as his height and weight. Both these variables are subject to considerable random fluctuation. Nevertheless we expect to find some relationship between them as a man who is taller than average is also likely to be heavier than average.

The problem in this sort of situation is to see if the two variables are inter-related, and, if so, to find a measure of the degree of association or *correlation* between them. We will only be concerned with linear correlation, when the relationship between the variables appears to be linear.

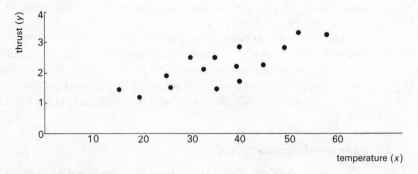

Figure 45   Scatter diagram of data of Example 5

The next problem is to standardize the covariance in such a way that it does not depend on the scales in which the measurements were made. Thus in Example 5 we would like to get the same measure of correlation if the temperature is measured in degrees Fahrenheit, centigrade, or any other scale. A convenient way of doing this is to express $(x_i - \bar{x})$ and $(y_i - \bar{y})$ in units of their respective standard deviations, s_x and s_y, where

$$s_x = \sqrt{\left[\frac{\sum(x_i - \bar{x})^2}{n-1}\right]} \quad \text{and} \quad s_y = \sqrt{\left[\frac{\sum(y_i - \bar{y})^2}{n-1}\right]}.$$

Then the required measure of correlation is given by

$$r = \frac{1}{n-1} \sum_{i=1}^{n} \left(\frac{x_i - \bar{x}}{s_x}\right)\left(\frac{y_i - \bar{y}}{s_y}\right),$$

which can be rewritten as Equation **8.5**

If a desk calculator is available it is usually more convenient to use the equivalent formula

$$r = \frac{\sum x_i y_i - \dfrac{(\sum x_i)(\sum y_i)}{n}}{\sqrt{\left[\left(\sum x_i^2 - \dfrac{(\sum x_i)^2}{n}\right)\left(\sum y_i^2 - \dfrac{(\sum y_i)^2}{n}\right)\right]}}$$

It can be shown that the value of r must lie between -1 and $+1$. For $r = +1$, all the observed points lie on a straight line which has a positive slope; for $r = -1$, all the observed points lie on a straight line which has a negative slope.

Example 6

Calculate the correlation coefficient of the data from Example 5.

$$n = 15, \quad \sum xy = 1276 \cdot 1, \quad \sum x = 540, \quad \sum y = 33,$$

$$\sum x^2 = 21426, \quad \sum y^2 = 78 \cdot 44.$$

Hence

$$r = \frac{88 \cdot 1}{\sqrt{(1986 \times 5 \cdot 84)}} = 0 \cdot 82$$

Thus the correlation is high and positive; this was to be expected after inspecting Figure 47.

It is often useful to perform a significance test to see if the observed correlation coefficient is significantly different from zero. If there is really no correlation between the two variables, it is still possible that a spuriously high (positive or negative) sample correlation value may occur by chance. When the true correlation coefficient is zero, it can be shown that the statistic $r\sqrt{(n-2)}/\sqrt{(1-r^2)}$ has a t-distribution with $n-2$ degrees of freedom, provided that both variables are normally distributed. If we are interested in positive or negative correlation then a two-tailed test is appropriate. The correlation is significantly different from zero at the α level of significance if

$$\left| \frac{r\sqrt{(n-2)}}{\sqrt{(1-r^2)}} \right| \geqslant t_{\alpha/2, n-2}.$$

Example 7

Is the correlation coefficient in Example 6 significantly different from zero?

$$\frac{r\sqrt{(n-2)}}{\sqrt{(1-r^2)}} = 9 \cdot 0$$

But $t_{0 \cdot 025, 13} = 2 \cdot 16$.

Thus the correlation between temperature and thrust is significant at the 5 per cent level.

Table 14 gives the 95 per cent critical points for the absolute value of the correlation coefficient for different sample sizes. When the sample size is small a fairly large absolute value of r is required to show significant correlation.

Table 14

Sample size	Critical value	Sample size	Critical value
5	0·75	25	0·38
10	0·58	30	0·35
15	0·48	50	0·27
20	0·42	100	0·20

Confidence intervals for the true correlation coefficient are given in Pearson and Hartley (1966).

8.10 Estimating the regression lines

If the size of the sample correlation coefficient indicates that the random variables are interdependent, then we may want to predict the value of one of the variables from a given value of the other variable. In this situation it is important to realise that there are two regression lines, one to predict y from x and one to predict x from y.

If the variables are linearly related, then the regression line of y on x can be denoted by

$$y = a_0 + a_1 x.$$

Estimates of a_0 and a_1 can be obtained by using the method of least squares, as described in section 8.2. We find

$$\hat{a}_0 = \bar{y} - \hat{a}_1 \bar{x},$$

$$\hat{a}_1 = \frac{\sum (x_i - \bar{x})(y_i - \bar{y})}{\sum (x_i - \bar{x})^2}.$$

We can denote the regression line of x on y by

$$x = b_0 + b_1 x.$$

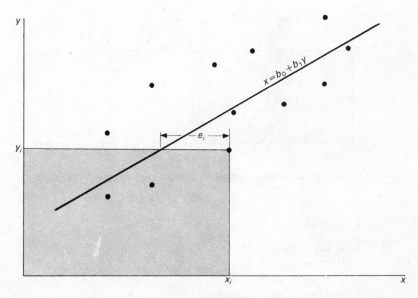

Figure 48   Finding the regression line of x on y

This line will generally be different from the regression line of y on x. Estimates of b_0 and b_1 can again be obtained by the method of least squares, but in this case they are obtained by minimizing the sum of squared deviations parallel to the x-axis and not parallel to the y-axis as in previous regression problems.

For any point (x_i, y_i) the deviation is given by

$$e_i = x_i - b_0 - b_1 y_i.$$

Minimizing $S = \sum e_i^2$ with respect to b_0 and b_1, we find

$$\hat{b}_0 = \bar{x} - \hat{b}_1 \bar{y},$$

$$\hat{b}_1 = \frac{\sum(x_i - \bar{x})(y_i - \bar{y})}{\sum(y_i - \bar{y})^2}.$$

The two regression lines will only coincide when the observations all lie on a straight line. Both lines pass through the centroid of the data, but their slopes are different. If there is no correlation between the variables then the two regression lines will be at right angles. In this case the best prediction of y for any value of x is simply $\bar{y}$. Conversely the best prediction of x for any value of y is simply $\bar{x}$.

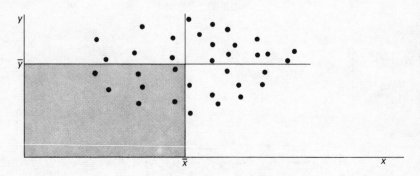

Figure 49 Regression lines when there is no correlation

Example 8

Find the regression lines for the data of Example 5.

We have $\bar{x} = 36$, $\bar{y} = 2\cdot2$.

$$\sum xy - \frac{(\sum x)(\sum y)}{15} = 88\cdot1,$$

$$\sum x^2 - \frac{(\sum x)^2}{15} = 1986,$$

$$\sum y^2 - \frac{(\sum y)^2}{15} = 5 \cdot 84.$$

Thus the estimated regression line of y on x is given by

$$\hat{y} - 2 \cdot 2 = \frac{88 \cdot 1}{1986}(x - 36)$$

$$= 0 \cdot 044(x - 36).$$

The estimated regression line of x on y is given by

$$\hat{x} - 36 = \frac{88 \cdot 1}{5 \cdot 84}(y - 2 \cdot 2)$$

$$= 15 \cdot 1(y - 2 \cdot 2).$$

These regression lines are plotted in Figure 50.

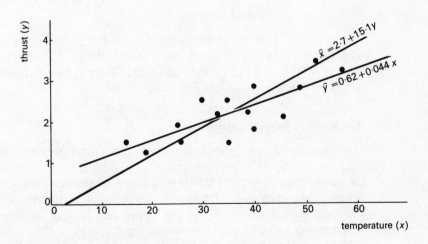

Figure 50   Regression lines for the data of Example 5

We are now in a position to show that the square of the correlation coefficient, r, is equal to the coefficient of determination which was discussed in section 8.5. The estimated regression line of y on x is given by

$$\hat{y} - y = \left[\frac{\sum(x_i - \bar{x})(y_i - \bar{y})}{\sum(x_i - \bar{x})^2} \right](x - \bar{x})$$

$$= r\frac{s_y}{s_x}(x - \bar{x}) \quad \text{using equation 8.5.}$$

Thus

$$\sum(\hat{y}_i - \bar{y})^2 = r^2\frac{s_y^2}{s_x^2}\sum(x_i - \bar{x})^2$$

$$= r^2 s_y^2(n-1)$$

$$= r^2 \sum(y_i - \bar{y})^2.$$

But

$$\sum(\hat{y}_i - \bar{y})^2 = \text{('explained variation')}$$

and

$$\sum(y_i - \bar{y})^2 = \text{('total variation')}.$$

Thus

$$r^2 = \text{(coefficient of determination)}.$$

This quantity gives the proportion of the total variation in y which is accounted for by the linear variation with x. Thus in Example 6 we find that $0.82^2 = 0.672$ of the total variation in thrust is accounted for by the linear relationship with temperature.

8.11 The bivariate normal distribution

In section 8.9 no assumptions were made about the distribution of the random variables. For many purposes it is convenient to assume that the pairs of measurements follow a *bivariate normal distribution*. This is the natural extension of the normal distribution to the case of two variables, and describes fairly well many distributions of pairs of measurements which arise in practice.

The formula for the joint probability density function of this distribution is rather complicated. Denote the two random variables by X and

Y. If a pair of observations is taken at random from a bivariate normal distribution, the probability that the value of X lies between x and $x + dx$ and that the value of Y lies between y and $y + dy$ is given by

$$f(x, y)\, dx\, dy = \frac{1}{2\pi\sigma_x\sigma_y\sqrt{(1-\rho^2)}} \times$$

$$\times \exp\left\{-\frac{1}{2(1-\rho^2)}\left[\left(\frac{x-\mu_x}{\sigma_x}\right)^2 - 2\rho\left(\frac{x-\mu_x}{\sigma_x}\right)\left(\frac{y-\mu_y}{\sigma_y}\right) + \left(\frac{y-\mu_y}{\sigma_y}\right)^2\right]\right\} dx\, dy,$$

where μ_x, σ_x, μ_y, σ_y denote the mean and standard deviation of X and Y respectively. The parameter ρ is the *theoretical correlation coefficient* and is defined by

$$\rho = E\left[\left(\frac{X-\mu_x}{\sigma_x}\right)\left(\frac{Y-\mu_y}{\sigma_y}\right)\right].$$

It is easy to show that ρ always lies between -1 and $+1$, and that it is zero when X and Y are independent. The sample correlation coefficient, r, introduced in section 8.9 is a point estimate of ρ, and Example 7 shows how to test the hypothesis $H_0 : \rho = 0$.

Given particular values for the five parameters, μ_x, σ_x, μ_y, σ_y and ρ, we can compute $f(x, y)$ at all values of (x, y). This will give a three-dimensional surface which has a maximum at (μ_x, μ_y) and which decreases down to zero in all directions. The total volume under the surface is one. The points (x, y) for which $f(x, y)$ is a constant form an ellipse with centre at (μ_x, μ_y). The major axes of these ellipses have a positive slope when ρ is positive and a negative slope when ρ is negative. In the special case when $\rho = 0$ the major axes of the ellipses are parallel to the x-axis if $\sigma_x > \sigma_y$ and parallel to the y-axis if $\sigma_x < \sigma_y$.

An important property of the bivariate normal distribution is that the conditional distribution of Y, for a given value x of X, is normal with mean

$$\mu_{y|x} = \mu_y + \rho\frac{\sigma_y}{\sigma_x}(x - \mu_x)$$

and variance

$$\sigma_{y|x}^2 = \sigma_y^2(1 - \rho^2).$$

This important result, which will not be proved here, says that the conditional means of the y-distributions lie on the straight line

$$y = \mu_y + \rho\frac{\sigma_y}{\sigma_x}(x - \mu_x),$$

which goes through the point (μ_x, μ_y) and which has a slope of $\rho \sigma_x / \sigma_y$. Thus the regression curve of y on x is a straight line.

This regression line can be estimated by the method of least squares. However since $\sigma^2_{y|x}$ is constant, these estimates will in fact be maximum likelihood estimates. The least squares estimate of the slope of the regression line is given by

$$\frac{\sum(x_i - \bar{x})(y_i - \bar{y})}{\sum(x_i - \bar{x})^2} \qquad \text{see section 8.2}$$

$$= r\frac{s_y}{s_x}. \qquad \text{from equation 8.5}$$

This quantity is in fact the intuitive estimate of $\rho \sigma_x / \sigma_y$, obtained by substituting the sample estimates of ρ, σ_x and σ_y.

For a bivariate normal distribution it can also be shown that the conditional distribution of X, for a given value y of Y, is normal with mean

$$\mu_{x|y} = \mu_x + \rho \frac{\sigma_x}{\sigma_y}(y - \mu_y).$$

Thus the regression curve of x on y is also a straight line, and the least squares estimates of the parameters of this line can again be obtained by substituting the sample values of μ_x, μ_y, σ_x, σ_y and ρ.

8.12 Interpretation of the correlation coefficient

We will conclude this chapter with a few general remarks about the correlation coefficient. Firstly it is worth emphasizing again that it

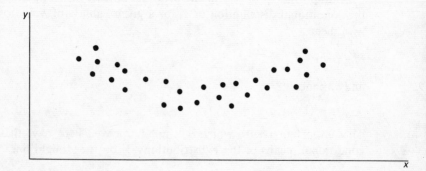

Figure 51   Correlation coefficient should not be calculated

should only be calculated when the relationship between two random variables is thought to be linear. If the scatter diagram indicates a non-linear relationship, as in Figure 51, then the correlation coefficient will be misleading and should not be calculated. The data in Figure 51 would give a value of r close to zero even though the variables are clearly dependent.

It is also important to realise that a high correlation coefficient between two variables does not necessarily indicate a causal relationship. There may be a third variable which is causing the simultaneous change in the first two variables, and which produces a spuriously high correlation coefficient. In order to establish a causal relationship it is necessary to run a carefully controlled experiment. Unfortunately it is often impossible to control all the variables which could possibly be relevant to a particular experiment, so that the experimenter should always be on the lookout for spurious correlation (see also section 10.4).

Exercises

1. The following measurements of the specific heat of a certain chemical were made in order to investigate the variation in specific heat with temperature.

Temperature °C	0	10	20	30	40
Specific heat	0·51	0·55	0·57	0·59	0·63

Plot the points on a scatter diagram and verify that the relationship is approximately linear. Estimate the regression line of specific heat on temperature, and hence estimate the value of the specific heat when the temperature is 25°C.

2. When the values of the controlled variable are equally spaced, the calculation of the regression line can be considerably simplified by coding the data in integers symmetrically about zero. When the values of the controlled variable sum to zero, the estimates of the regression coefficients are given by

$$\hat{a}_0 = \bar{y}, \qquad \hat{a}_1 = \frac{\sum x_i y_i}{\sum x_i^2}.$$

For example, with five equally spaced values of the controlled variable the coded values could be $-2, -1, 0, +1, +2$; with six values the coded measurements could be $-5, -3, -1, +1, 3, 5$.

Use this technique to find the regression line in the following case.

Output (1000 tons)	11·1	12·3	13·7	14·6	15·6
Year	1960	1961	1962	1963	1964

Estimate the output of the company in 1965.

3. The following are the measurements of the height and weight of ten men.

Height (inches)	63	71	72	68	75	66	68	76	71	70
Weight (pounds)	145	158	156	148	163	155	153	158	150	154

(a) Calculate the correlation coefficient and show that it is significantly different from zero.

(b) Find the linear regression of height on weight.

(c) Find the linear regression of weight on height.

4. In this example we have coded measurements on a dependent variable y, and two controlled variables x_1 and x_2.

Test	y	x_1	x_2
1	1·6	1	1
2	2·1	1	2
3	2·4	2	1
4	2·8	2	2
5	3·6	2	3
6	3·8	3	2
7	4·3	2	4
8	4·9	4	2
9	5·7	4	3
10	5·0	3	4

Find the linear regression of y on x_1 and x_2.

References

FISHER, R. A., and YATES, F. (1963), *Statistical Tables for Biological, Agricultural and Medical Research*, Oliver & Boyd, 6th edn.

HALD, A. (1952), *Statistical Theory with Engineering Applications*, Wiley.

PEARSON, E. S., and HARTLEY, H. O. (1966), *Biometrika Tables for Statisticians*, Cambridge University Press, 3rd edn.

199 **Exercises**

Part three
Applications

He had forty-two boxes, all carefully packed,
With his name painted clearly on each:
But, since he omitted to mention the fact,
They were all left behind on the beach.

Chapter 9
Planning the experiment

The next three chapters will be concerned with the design and analysis of experiments. This chapter, entitled the planning of experiments, deals with preliminary considerations, particularly that of finding the precision of the response variable. Chapter 10 covers the design and analysis of comparative experiments, while Chapter 11 deals with the design and analysis of factorial experiments.

9.1 Preliminary remarks

Probably more experimental effort is wasted because problems have been poorly defined than for any other reason. When designing an experiment the first task of the engineer or scientist is to study the physical aspects of the experiment to the point where he is confident that there is no relevant knowledge of which he is not aware. The next task is to make a precise definition of the problem. Many people underestimate the time required to accomplish these two tasks. Experience suggests that although the final experimental programme may be completed in less than a month, it may take as long as a year to reach an understanding of the problem and false starts are to be expected. The engineer should always be prepared to do more research, more analysis and more literary search before plunging into a poorly defined test programme.

Furthermore a scientist with little knowledge of statistics should not attempt to design an experiment by himself but should call in the services of a statistician. In industry the technologist who has studied statistics with some thoroughness will often act as consultant to other engineers and scientists. The following remarks are addressed to such a statistician. Firstly he should make it quite clear that his objective is to assist the experimenter and not replace him. Secondly he should be prepared to act as a restraining influence. When an experimental result is disappointing, the new invention malfunctions, or an attractive hypothesis is rejected by a significance test, there is an urgent desire to 'test something', to 'get going'. The statistician must always be prepared to encourage the experimenter to make more thorough preparations. Thirdly there is the problem of what to do with prior data. The statistician is often consulted only after considerable testing has been completed in vain without solving the problem. Although this data may be worthless due to the absence of good statistical procedure, it may contain some information and it would be stupid to discard it without a glance. An inspection may reveal that some worthwhile analysis has not been done.

9.2 Measurements

Once the problem has been well defined the next problem is to determine the quantities to be measured. The response variables in which we are really interested cannot always be measured directly but may be a function of the measurements. These measurements are basic quantities such as length, weight, temperature or the resistance of a thermo-couple. But the response variables may be more complex quantities such as efficiency, fuel consumption per unit of output etc. If the functional relationship between the response variable and the measurements is known, the measurements can be transformed by this function to obtain the required quantities.

Unfortunately the measurements are almost certain to contain an error component. By now you should be convinced that no two experiments are identical except in an approximate sense. There are an infinite number of reasons why small variations will occur in repeated tests. The power supply changes, the technician is tired, weather conditions vary; and so on. If the variation is small the experiment is said to be *precise* or repeatable. Precision is expressed quantitatively as the standard deviation of results from repeated trials under identical conditions. This precision is often known from past experiments, however if it is not known then one of the experimental test objectives should be to obtain an estimate of it.

Another useful property of 'good' measurements is that they should be *accurate* or free from bias. An experiment is said to be accurate or unbiased if the expected value of the measurement is equal to the true value. In other words there is no systematic error. Measuring instruments should be checked for accuracy by relating them to some defined standard, which is accepted, for example, by the National Bureau of Standards in the U.S.A.

If the measured values are plotted against the standard reference values, a *calibration curve* results. This curve enables us to estimate the true value by multiplying the measured value by a correction factor. Calibration problems are discussed in Mandel (1964). It may not be possible to entirely eliminate all bias but it can often be made as small as we wish. It is customary to estimate the bias of an instrument in the form of an upper bound (not greater than —). This is often an experienced guess.

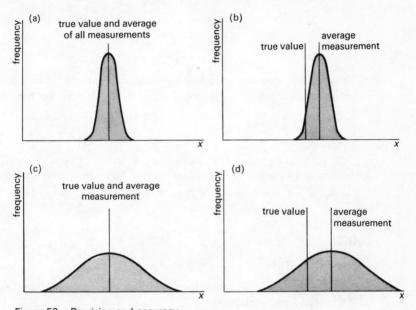

Figure 52   Precision and accuracy
(a) accurate and precise (b) not accurate but precise
(c) accurate but not precise (d) not accurate, not precise

It is clear from the above that 'good' measurements have the same properties as 'good' estimators, namely precision and accuracy (see section 6.5). Four types of distributions of measurements are illustrated in Figure 52. Because there is variation in the measurements, there will also be variation in the corresponding values of the response variable. The effect of the functional relationship with the measurements on the error in the response variable is sometimes called the *propagation of error*. The size of the error in the response variable will depend on the precision and accuracy of the measurements. In the following sections we will usually make the assumption that the measurements are accurate. This is reasonable provided that all meters are recalibrated at regular intervals.

9.3 The propagation of error

We have seen that the response variable is often some function of the observed measured variables. Our primary concern is to find the distribution of the response variable so that, for example, we can judge if the variation is unacceptably large. This section will be concerned with methods for finding the distribution of the response variable.

9.3.1 *Linear functions*

The simplest case occurs when the response variable z, is the sum of several independent measured variables which will be denoted by $x_1, x_2, \ldots, x_n$.

$$z = x_1 + x_2 + \ldots + x_n.$$

Denote the true values of these measured variables by $\mu_1, \mu_2, \ldots, \mu_n$. If the measurements are accurate and have respective precisions $\sigma_1, \sigma_2, \ldots, \sigma_n$ then, using the results of section 6.2, we have

$$E(z) = \mu_1 + \mu_2 + \ldots + \mu_n = \mu_z$$

and

$$\text{variance } (z) = \sigma_1^2 + \sigma_2^2 + \ldots + \sigma_n^2 = \sigma_z^2.$$

For the general case, where the response variable is any linear function of the measured variables,

$$z = a_1 x_1 + a_2 x_2 + \ldots + a_n x_n,$$

where the a_is are constants. Then we have

$$E(z) = a_1 \mu_1 + a_2 \mu_2 + \ldots + a_n \mu_n = \mu_z$$

and

$$\text{variance } (z) = a_1^2 \sigma_1^2 + a_2^2 \sigma_2^2 + \ldots + a_n^2 \sigma_n^2 = \sigma_z^2.$$

Thus having found the mean and variance of the distribution of z, the next question that arises is to find the type of distribution. The following important result is stated without proof. If the response variable is a linear function of several measured variables, each of which is normally distributed, then the response variable will also be normally distributed. So given observations on the measured variables, the corresponding value of the response variable can be found. This value, denoted by z_0, is a point estimate of μ_z. Since the response variable is normally distributed, the $100(1-\alpha)$ per cent confidence interval for μ_z is given by $z_0 \pm z_{\alpha/2}\sigma_z$, where $z_{\alpha/2}$ is the appropriate percentage point of the standard normal distribution.

Tolerances. One important application of the above results is in the study of tolerances, where the dimensions of manufactured products have to be carefully controlled. The *tolerance limits* of a particular dimension are defined to be those values between which nearly all the manufactured items will lie. If measurements on this dimension are found to be normally distributed with mean μ and precision σ, then the tolerance limits are usually taken to be $\mu \pm 3\sigma$ (see also section 12.9). Then only $0 \cdot 27$ per cent of all items can be expected to fall outside the tolerance limits.

It often happens that a product is made by assembling several parts and so one dimension of the product may be the sum of the dimensions of the constituent parts. In this case the above formulae can be used to find the tolerance limits of the dimension of the product.

Example 1

An item is made by adding together three components whose lengths are x_1, x_2 and x_3. The over-all length is denoted by z.

$$z = x_1 + x_2 + x_3.$$

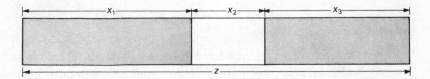

Figure 53

The tolerance limits (in inches) of the lengths of the three components are known to be $1 \cdot 960 \pm 0 \cdot 030$, $0 \cdot 860 \pm 0 \cdot 030$ and $1 \cdot 865 \pm 0 \cdot 015$, respectively. Thus the respective precisions of x_1, x_2, and x_3 are $0 \cdot 010$, $0 \cdot 010$, and $0 \cdot 005$ in.

If the lengths of the three components are independently normally distributed, then the over-all length is also normally distributed with

$$E(z) = 1.960 + 0.860 + 1.865$$

$$= 4.685$$

and

$$\text{variance } (z) = \sigma_z^2 = 0.01^2 + 0.01^2 + 0.005^2$$

$$= 0.000225,$$

giving $\sigma_z = 0.015$.

Thus the tolerance limits for the over-all length are 4.685 ± 0.045.

Example 2

Suppose that the tolerance limits for the over-all length, which were calculated in Example 1, were judged to be too wide. If the tolerance limits for x_2 and x_3 cannot be narrowed, find reduced tolerance limits for x_1 so that the over-all tolerance limits are 4.685 ± 0.036.

Denote the revised precision of x_1 by σ_1. The revised precision of the over-all length is given by 0.012. Thus we have

$$0.012^2 = \sigma_1^2 + 0.01^1 + 0.005^2,$$

giving $\sigma_1 = 0.0044$.

The revised tolerance limits for x_1 are 1.960 ± 0.020. This, of course, means that a considerable improvement must be made in the manufacturing standards of this component in order to achieve the required precision.

9.3.2 *Non-linear functions*

Thus far we have considered the case where the response variable is a linear function of the measured variables. We now turn our attention to non-linear functions. For simplicity let us begin by assuming that the response variable, z, is a known function of just one measured variable, x, that is, $z = f(x)$, and let us denote the true value of x by μ. If the measurements are accurate and have precision σ, then successive measurements will form a distribution with mean μ and standard deviation σ. This distribution is sometimes called the error distribution of x.

The true value of the response variable is given by $\mu_z = f(\mu)$. If all the measurements are transformed by the given function, then these transformed values form the error distribution of the response variable. We would like to answer three questions about this distribution. Firstly we would like to know if the mean value of this distribution is equal to μ_z. In other words given a measurement x, we would like to know if $f(x)$ is an unbiased estimate of $f(\mu)$. Secondly we would like to find the standard deviation of this distribution. In other words we would like to find the precision of the response variable. Thirdly if x is normally distributed, we would like to find the distribution of z. These questions can be answered in the following way.

Provided that $f(x)$ is a continuous function near μ, it can be expanded as a Taylor series about the point $x = \mu$.

$$f(x) = f(\mu) + \left(\frac{df}{dx}\right)_\mu (x-\mu) + \frac{1}{2}\left(\frac{d^2f}{dx^2}\right)_\mu (x-\mu)^2 + R_3,$$

where df/dx and d^2f/dx^2 are evaluated at $x = \mu$. R_3 is the remainder term which can be disregarded if higher derivatives are small or zero and the coefficient of variation of x is reasonably small.

The above equation can be rewritten in the equivalent form

$$z \approx \mu_z + \left(\frac{dz}{dx}\right)_\mu (x-\mu) + \frac{1}{2}\left(\frac{d^2z}{dx^2}\right)_\mu (x-\mu)^2.$$

Then
$$E(z) \approx E\left[\mu_z + \left(\frac{dz}{dx}\right)_\mu (x-\mu) + \frac{1}{2}\left(\frac{d^2z}{dx^2}\right)_\mu (x-\mu)^2\right]$$

$$= \mu_z + \left(\frac{dz}{dx}\right)_\mu E(x-\mu) + \frac{1}{2}\left(\frac{d^2z}{dx^2}\right)_\mu E[(x-\mu)^2]$$

But

$$E(x-\mu) = 0$$

and

$$E[(x-\mu)^2] = \sigma^2,$$

thus

$$E(z) \approx \mu_z + \frac{1}{2}\left(\frac{d^2z}{dx^2}\right)_\mu \sigma^2.$$

209 The propagation of error

The expression $\frac{1}{2}(d^2z/dx^2)_\mu \sigma^2$ is the bias resulting from using $f(x)$ as an estimate of $f(\mu)$. Fortunately this bias term will often be small or zero in which case we have

$$E[f(x)] \approx f(\mu)$$

or

$$E(z) \approx \mu_z.$$

Example 3

The measurement x, which is an unbiased estimate of μ, is transformed by the equation $z = \sqrt{x}$. Find the bias in using z as an estimate of $\sqrt{\mu}$.

$$\frac{dz}{dx} = \frac{1}{2\sqrt{x}} \qquad \frac{d^2z}{dx^2} = -\frac{1}{4x^{3/2}}.$$

Thus $\quad E(z) \approx \mu_z - \dfrac{\sigma^2}{8\mu^{3/2}} \quad$ where $\quad \mu_z = \sqrt{\mu}.$

Hence $\quad \dfrac{E(z)}{\mu_z} \approx 1 - \dfrac{\sigma^2}{8\mu^2}.$

Thus the percentage bias in estimating $\sqrt{\mu}$ can be calculated for different values of the coefficient of variation of x, σ/μ.

$\dfrac{\sigma}{\mu}$	Percentage bias
0·1	0·125%
0·2	0·5%
0·3	1·125%

Thus even when the coefficient of variation of x is as high as 30 per cent, the percentage bias in the transformed values is only 1·125 per cent.

In the remainder of this section, we will assume that the bias term, $\frac{1}{2}(d^2z/dx^2)_\mu \sigma^2$ is relatively small, so that we can write

$$z \approx \mu_z + \left(\frac{dz}{dx}\right)_\mu (x - \mu).$$

The important point to notice here is that this is now a linear function of x. In other words we have approximated $f(x)$ by a linear function in the area of interest. This means that if x is normally distributed, then z will also be approximately normally distributed. Furthermore the precision of z can be found as follows.

Since $\quad E(z) \approx \mu_z$

and $\quad z - \mu_z \approx \left(\dfrac{dz}{dx}\right)_\mu (x - \mu),$

variance $(z) = \sigma_z^2 \approx E[(z - \mu_z)^2]$

$$\approx E\left[\left(\dfrac{dz}{dx}\right)_\mu^2 (x - \mu)^2\right]$$

$$= \left(\dfrac{dz}{dx}\right)_\mu^2 \sigma^2.$$

Hence $\sigma_z =$ (precision of the response variable)

$$\approx \left(\dfrac{dz}{dx}\right)_\mu \sigma.$$

Example 4

It is instructive to check the above formula when $f(x)$ is a linear function.

Let $\quad z = ax + b,$

then $\quad \dfrac{dz}{dx} = a.$

Therefore variance $(z) = a^2\sigma^2$; this was the result obtained earlier.

Example 5

The measurement x, which is an unbiased estimate of μ, is transformed by the equation $z = \sqrt{x}$. Find the coefficient of variation of z in terms of the coefficient of variation of x.

$$\left(\dfrac{dz}{dx}\right)_\mu = \dfrac{1}{2\sqrt{\mu}}$$

variance $(z) \approx \dfrac{\sigma^2}{4\mu}$ where σ is the precision of x.

$$\sigma_z \approx \frac{\sigma}{2\sqrt{\mu}}$$

$$\text{(coefficient of variation of } z) = \frac{\sigma_z}{\mu_z}$$

$$= \frac{\sigma/2\sqrt{\mu}}{\sqrt{\mu}}$$

$$= \frac{\sigma}{2\mu}$$

$$= \tfrac{1}{2}\text{(coefficient of variation of } x).$$

It often happens that several measurements are made on the measured variable. Then it is better to find the average measurement, $\bar{x}$, and transform this value, rather than transforming each measurement and finding the average of the transformed measurements. This procedure will reduce the bias in the estimation of the true value of the response variable. If there are n observations, the precision of $\bar{x}$ is given by $\sigma/\sqrt{n}$ and then the precision of the response variable will be given by $(df/d\bar{x})_\mu \, \sigma/\sqrt{n}$.

Example 6

Four measurements, 4·01, 4·08, 4·14 and 4·09 are made on a measured variable whose unknown true value is denoted by μ. These measurements are known to be accurate and to have precision $\sigma = 0·06$. Find a 95 per cent confidence interval for $\sqrt{\mu}$.

The observed average measurement, denoted by $\bar{x}_0$, is equal to 4·08. Thus a point estimate of $\sqrt{\mu}$ is given by

$$\sqrt{\bar{x}_0} = 2·02.$$

The precision of this estimate depends on $(dz/d\bar{x})_\mu$, where $z = \sqrt{\bar{x}}$. Unfortunately this cannot be evaluated exactly as we do not know the true value of μ. However it is sufficiently accurate to evaluate $(dz/d\bar{x})$ at the point $\bar{x}_0$. Thus the precision of the estimate of $\sqrt{\mu}$ is given by

$$\sigma_z \approx \left(\frac{dz}{d\bar{x}}\right)_{\bar{x}_0} \sigma_{\bar{x}}$$

$$= \frac{\sigma}{2\sqrt{(n\bar{x}_0)}}$$

$$= 0·0075.$$

Thus a 95 per cent confidence interval for $\sqrt{\mu}$ is given by $2 \cdot 02 \pm 1 \cdot 96 \times 0 \cdot 0075$.

More generally, we can consider the situation where the response variable, z, is a function of several independent measured variables, $x_1, x_2, \ldots, x_n$.

$$z = f(x_1, x_2, \ldots, x_n).$$

Let $\mu_1, \mu_2, \ldots, \mu_n$ denote the true values of the measured variables. Assume that all the measurements are accurate and have precisions $\sigma_1, \sigma_2, \ldots, \sigma_n$, respectively. The true value of z is given by $\mu_z = f(\mu_1, \mu_2, \ldots, \mu_n)$. As before, we can expand $f(x_1, x_2, \ldots, x_n)$ in a Taylor series about the point $(\mu_1, \mu_2, \ldots, \mu_n)$. If the measured variables have fairly small coefficients of variation, and if the second order partial derivatives $(\partial^2 z / \partial x_i^2)$ are small or zero at $(\mu_1, \mu_2, \ldots, \mu_n)$, then we find

$$E(z) \approx \mu_z$$

and $\quad \sigma_z^2 \approx \left(\dfrac{\partial z}{\partial x_1} \right)^2 \sigma_1^2 + \left(\dfrac{\partial z}{\partial x_2} \right)^2 \sigma_2^2 + \ldots + \left(\dfrac{\partial z}{\partial x_n} \right)^2 \sigma_n^2,$

where the partial derivatives are evaluated at $(\mu_1, \ldots, \mu_n)$. When these true values are unknown it is usually sufficiently accurate to estimate σ_z^2 by evaluating the partial derivatives at the observed values of the measured variables.

The above formulae are strictly accurate only for linear functions, but can also be used for functions involving products and quotients provided that the coefficients of variation of the measured variables are less than about 15 per cent.

Example 7

Assume that x and y are independent measured variables with mean μ_x, μ_y, and precision σ_x, σ_y, respectively. Also assume that σ_x/μ_x and σ_y/μ_y are less than about $0 \cdot 15$. Calculate the expected value and precision of the response variable, z, in the following cases:

(a) *A linear function*

$z = ax + by$—where a, b are positive or negative constants.

$$\frac{\partial z}{\partial x} = a \qquad \frac{\partial z}{\partial y} = b.$$

All higher partial derivatives are zero.

Hence $\quad E(z) = a\mu_x + b\mu_y,$

variance $(z) = a^2\sigma_x^2 + b^2\sigma_y^2.$

These results are the same as those obtained earlier.

(b) *Products*

$z = xy,$

$E(z) \approx \mu_z = \mu_x.\mu_y.$

$$\frac{\partial z}{\partial x} = y \qquad \frac{\partial z}{\partial y} = x.$$

If these are evaluated at (μ_x, μ_y), we find

$$\sigma_z^2 \approx \mu_y^2\sigma_x^2 + \mu_x^2\sigma_y^2,$$

$$\frac{\sigma_z^2}{\mu_z^2} \approx \frac{\sigma_x^2}{\mu_x^2} + \frac{\sigma_y^2}{\mu_y^2}$$

or

$$C_z^2 \approx C_x^2 + C_y^2,$$

where C_x, C_y, C_z are the coefficients of variation of x, y and z respectively.

(c) *Quotients*

$$z = \frac{x}{y},$$

$$E(z) \approx \mu_z = \frac{\mu_x}{\mu_y}.$$

$$\frac{\partial z}{\partial x} = \frac{1}{y} \qquad \frac{\partial z}{\partial y} = -\frac{x}{y^2}.$$

If these are evaluated at (μ_x, μ_y) we find

$$\sigma_z^2 \approx \frac{\sigma_x^2}{\mu_y^2} + \sigma_y^2 \times \left(\frac{-\mu_x}{\mu_y^2}\right)^2,$$

$$\frac{\sigma_z^2}{\mu_z^2} \approx \frac{\sigma_x^2}{\mu_x^2} + \frac{\sigma_y^2}{\mu_y^2}$$

or $\quad C_z^2 \approx C_x^2 + C_y^2.$

We are now in a position to find the precision of the response variable if we are given the precision of the measured variables. The latter are usually known from past experiments; but if they are not known then they should be estimated by taking repeated measurements under controlled conditions.

Should the response error be found to be unacceptably large there are several possible courses of action. Firstly an attempt could be made to improve the precision of the measured variables, either by improving the instrumentation or by improving the design of the system. Secondly it may be possible to use a different function to calculate the response from the measurements. Thirdly it may be possible to combine the information from several imprecise measurements to obtain an acceptable estimate of the response variable. This last method is possible when the system includes some redundant meters. The remainder of the chapter will be concerned with obtaining the most precise estimate of the response variable when this is the case.

9.4 Improving precision with series and parallel arrangements

A common way of improving the precision of a particular measurement is to add redundant meters in series and parallel arrangements. For example, suppose that we wanted to estimate the total flow through a pipe-line. Instead of just taking one observation, we could make three separate measurements, x_1, x_2 and x_3, on three meters in series, as shown in Figure 54. If the meters have the same precision σ and are unbiased, then the best unbiased estimate of the total flow is the average of the three measurements

$$z = \frac{x_1 + x_2 + x_3}{3}.$$

Figure 54   Meters in series

The precision of z is given by $\sigma/\sqrt{3}$, which is of course smaller than the precision of the individual measurements.

Alternatively, if there are different size meters available, or if the precision is a function of the flow through the meter, then a parallel arrangement, or a combination of meters in series and parallel, may be used. Two such arrangements are illustrated in Figure 55.

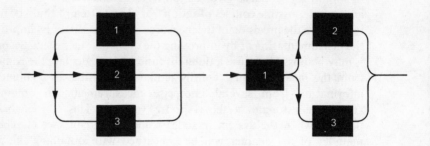

Figure 55　(a) Parallel (b) Series-parallel

For the parallel arrangement, the total flow is estimated by

$$z = x_1 + x_2 + x_3.$$

But for the series-parallel arrangement it is not clear what the best estimate of the total flow should be. Nor is it clear what the best estimate of the total flow would be in the series arrangement if the precisions of the three meters were not the same. Thus in the next section we will outline a general method for combining different measurements to obtain the best estimate of the response variable. This method can be used in the above type of situation or wherever information is available from several sources.

9.5　Combining dissimilar estimates by the method of least squares

We have seen that the true value of the response variable can often be estimated in several different ways using the information obtained from the different measuring instruments. A common procedure is to take the reading from the most precise instrument and use the redundant meters as checks. However we can combine all the information and obtain a single estimate by applying the method of least squares. This estimate will be better than the individual estimates because it will be more precise.

We have already described in Chapter 8 how to fit a straight line to data by the method of least squares. It is a relatively easy matter to adapt this method in order to estimate parameters in other situations. For example, suppose that n observations $x_1, x_2, \ldots, x_n$ are made on an unknown quantity, μ. We have

$$x_i = \mu + e_i \qquad (i = 1, \ldots, n),$$

where $e_1, e_2, \ldots, e_n$ are the measurement errors.

The sum of squared errors is given by

$$S = \sum_{i=1}^{n} (x_i - \mu)^2,$$

$$\frac{dS}{d\mu} = -2 \sum_{i=1}^{n} (x_i - \mu).$$

The least squares estimate of μ is obtained by minimizing S with respect to μ: this is achieved by putting $dS/d\mu = 0$. Then we find

$$\sum (x_i - \hat{\mu}) = 0,$$

$$\hat{\mu} = \sum \frac{x_i}{n} = \bar{x}.$$

Thus the sample mean is also the least squares estimate of μ.

The least squares method can be used whenever we have a set of measurements, each of which has an expected value which is a function of the unknown parameters. The sum of squared differences between the measurements and their expected values is formed, and then minimized with respect to the unknown parameters. The basic assumption necessary to apply the least squares method in this way is that the precision of the observations should be the same. Later on we shall see how to modify the method when the precisions are unequal. The procedure is best illustrated with an example.

Example 8

Suppose that we want to determine the flow in a system with one input and two output streams, all three being metered. This situation is similar to the series-parallel arrangements depicted in Figure 56.

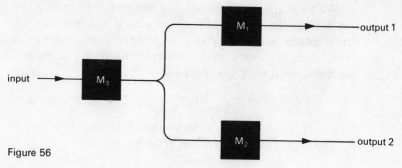

Figure 56

We have four pieces of information: the three meter readings, m_1, m_2 and m_3, all subject to precision error, plus the knowledge that the input is equal to the sum of the output streams. Let

$\mu_1 = $ (true flow in output 1),

$\mu_2 = $ (true flow in output 2),

then $\mu_1 + \mu_2 = $ (true input flow).

The relations between the true values and the measurements are given by

$$m_1 = \mu_1 + e_1,$$

$$m_2 = \mu_2 + e_2,$$

$$m_3 = \mu_1 + \mu_2 + e_3,$$

where e_1, e_2, e_3 are the measurement errors associated with the three meters. We begin by assuming that the meters have the same precision so that

$$\mathrm{var}(m_1) = \mathrm{var}(m_2) = \mathrm{var}(m_3) = \sigma^2.$$

The method of least squares can then be used to estimate μ_1 and μ_2.

For particular values of μ_1 and μ_2, the sum of squared errors is given by

$$S = (m_1 - \mu_1)^2 + (m_2 - \mu_2)^2 + (m_3 - \mu_1 - \mu_2)^2.$$

The least squares estimates of μ_1 and μ_2 are obtained by minimizing this expression with respect to μ_1 and μ_2.

$$\frac{\partial S}{\partial \mu_1} = -2(m_1 - \mu_1) - 2(m_3 - \mu_1 - \mu_2),$$

$$\frac{\partial S}{\partial \mu_2} = -2(m_2 - \mu_2) - 2(m_3 - \mu_1 - \mu_2).$$

Equating the two partial derivatives to zero and solving the two simultaneous equations, we get

$$\hat{\mu}_1 = \tfrac{1}{3}(2m_1 - m_2 + m_3),$$

$$\hat{\mu}_2 = \tfrac{1}{3}(2m_2 - m_1 + m_3),$$

$$\hat{\mu}_1 + \hat{\mu}_2 = \text{(least squares estimate of total flow)}$$

$$= \tfrac{1}{3}(m_1 + m_2 + 2m_3).$$

If we assume that the measurement errors are independent, then the precision of the least squares estimates can be found from

$$\sigma_{\hat{\mu}_1}^2 = \tfrac{1}{9}[4\sigma^2 + \sigma^2 + \sigma^2] = \frac{2\sigma^2}{3},$$

$$\sigma_{\hat{\mu}_2}^2 = \tfrac{1}{9}[4\sigma^2 + \sigma^2 + \sigma^2] = \frac{2\sigma^2}{3},$$

$$\sigma_{\hat{\mu}_3}^3 = \tfrac{1}{9}[\sigma^2 + \sigma^2 + 4\sigma^2] = \frac{2\sigma^2}{3}.$$

Thus, by using $\hat{\mu}_1$, $\hat{\mu}_2$, and $\hat{\mu}_1 + \hat{\mu}_2$ instead of m_1, m_2 and m_3 to estimate the respective flow rates, the precision has been improved from σ to $\sqrt{(2/3)}\sigma$ in each case.

9.5.1 Unequal precision errors

We have seen how to apply the method of least squares when the precision of the different observations is the same. However situations often arise where the instruments are not equally precise and then the method must be modified by giving more weight to the observations with the smaller precision.

Let us assume that readings are taken from n instruments and that σ_i is the precision of the ith instrument. Then a set of constants $P_1, P_2, \ldots, P_n$ must be chosen so that

$$\sigma_1^2 P_1 = \sigma_2^2 P_2 = \ldots = \sigma_n^2 P_n.$$

A convenient set of constants are defined by

$$P_i = \frac{\sigma_k^2}{\sigma_i^2} \quad (i = 1, \ldots, n),$$

where σ_k^2 is the square of the smallest precision. Thus P_k is the largest constant and is equal to one. As before the expected value of each measurement will depend on certain unknown parameters. In order to obtain the least squares estimates, the difference between each measurement and its expected value is squared, and then multiplied by the appropriate P constant. The adjusted sum of squares is formed by adding up these quantities, and the least squares estimates of the unknown parameters found by minimizing the adjusted sum of squares. The method is best illustrated with an example.

Example 9

Two chemists each make three measurements on the percentage of an important chemical in a certain material. The observations are x_{11}, x_{12}, x_{13} and x_{21}, x_{22}, x_{23}. Suppose that both chemists make unbiased measurements but that the variance of measurements made by the second chemist is four times the variance of measurements made by the first chemist. Find the least squares estimate of the true percentage of the chemical.

Denote the true percentage of the chemical by μ. Because the variances are unequal we do not take the simple average of the six observations. Instead we want to give more weight to the first three observations.

The constants P_{1i}, P_{2j} are given by

$$P_{1i} = 1 \quad (i = 1, 2, 3) \quad \text{and} \quad P_{2j} = \tfrac{1}{4} \quad (j = 1, 2, 3).$$

The adjusted sum of squared deviations is given by

$$S = \sum_{i=1}^{3} (x_{1i} - \mu)^2 + \frac{1}{4} \sum_{j=1}^{3} (x_{2j} - \mu)^2,$$

$$\frac{dS}{d\mu} = -2 \sum_{i=1}^{3} (x_{1i} - \mu) - \frac{2}{4} \sum_{j=1}^{3} (x_{2j} - \mu).$$

Putting this derivative equal to zero we obtain the least squares estimate of μ.

$$\hat{\mu} = \frac{1}{7}\left[4\sum_{i=1}^{3} x_{1i} + \sum_{j=1}^{3} x_{2j}\right].$$

Thus four times as much weight is given to the first three observations as to the other three observations.

Example 10

The system depicted in Figure 56 is similar to the system used to measure the flow of liquid oxygen through a rocket engine. There are three meters which measure the total flow and its two components, pre-burner flow and main burner flow. The precision of these three meters is known to be the following:

pre-burner flow (m_1):   2·85 gallons per minute

main burner flow (m_2):   20·37 gallons per minute

total flow (m_3):   25·82 gallons per minute.

Denote the true flows through M_1, M_2 and M_3 by μ_1, μ_2 and $\mu_1 + \mu_2$ respectively, and the meter readings by m_1, m_2, and m_3. Then the weighted sum of squares is given by

$$S = P_1(m_1 - \mu_1)^2 + P_2(m_2 - \mu_2)^2 + P_3(m_3 - \mu_1 - \mu_2)^2,$$

where

$$P_1 = \frac{(2\cdot85)^2}{(2\cdot85)^2} = 1\cdot0$$

$$P_2 = \frac{(2\cdot85)^2}{(20\cdot37)^2} = 0\cdot0196$$

$$P_3 = \frac{(2\cdot85)^2}{(25\cdot82)^2} = 0\cdot0122.$$

$$\frac{\partial S}{\partial \mu_1} = -2P_1(m_1 - \mu_1) - 2P_3(m_3 - \mu_1 - \mu_2),$$

$$\frac{\partial S}{\partial \mu_2} = -2P_2(m_2 - \mu_2) - 2P_3(m_3 - \mu_1 - \mu_2).$$

By equating these partial derivatives to zero we obtain two simultaneous equations in $\hat{\mu}_1$ and $\hat{\mu}_2$ which can be solved to give

$$\hat{\mu}_1 = \frac{m_1(P_1 P_2 + P_1 P_3) + P_2 P_3(m_3 - m_2)}{P_1 P_2 + P_1 P_3 + P_2 P_3},$$

$$\hat{\mu}_2 = \frac{m_2(P_1 P_2 + P_2 P_3) + P_1 P_3(m_3 - m_1)}{P_1 P_2 + P_1 P_3 + P_2 P_3}.$$

Note that if $P_1 = P_2 = P_3 = 1$, the precisions are equal and the equations are the same as those obtained in Example 8.

Substituting the values of P_1, P_2 and P_3, we find

$$\hat{\mu}_1 = 0.9925 m_1 + 0.0075(m_3 - m_2),$$

$$\hat{\mu}_2 = 0.6193 m_2 + 0.3808(m_3 - m_1),$$

$$\hat{\mu}_1 + \hat{\mu}_2 = 0.6118(m_1 + m_2) + 0.3883 m_3.$$

The precision of these three estimates turn out to be

$$\sigma_{\hat{\mu}_1} = 2.84,$$

$$\sigma_{\hat{\mu}_2} = 16.03,$$

$$\sigma_{\hat{\mu}_1 + \hat{\mu}_2} = 16.09.$$

The precision of the least squares estimate of μ_1 is only fractionally less than the precision of m_1, but the precision of the least squares estimates of μ_2 and $(\mu_1 + \mu_2)$ are substantially less than the precision of m_2 and m_3, and this makes the least squares procedure worthwhile.

Exercises

1. The measurement, x, is an unbiased estimate of μ and has precision σ. Expand x^2 in a Taylor series about the point $x = \mu$. Hence show that, if the coefficient of variation of x is small, the expected value of $z = x^2$ is given by $\mu^2 + \sigma^2$ and that the coefficient of variation of z is approximately twice as large as that of x.

2. The three measured variables x_1, x_2 and x_3 have small independent errors. The precisions are related by

$$\sigma_{x_1} = \sigma_{x_2} = 2\sigma_{x_3}.$$

The response variable, z, is given by

$$z = \frac{x_1 + x_2}{3x_3}.$$

Estimate the precision of z in terms of the precision of x_3 when the measurements are $x_1 = 2\cdot37$, $x_2 = 3\cdot1$ and $x_3 = 2\cdot08$.

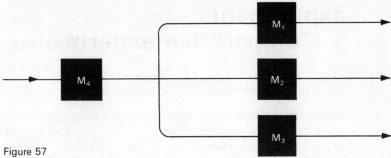

Figure 57

3. A system has one input and three output streams. Let m_1, m_2, m_3 and m_4 denote the reading of the respective meters. Assuming that the meters are unbiased and have equal precision, find the least squares estimate of the total flow.

4. Two flowmeters are connected in series. Let m_1, m_2, be the readings of meters A, B respectively. Assuming that both meters are unbiased and that the precision of meter A is $\pm 3\,\text{lb/hr}$ and that of meter B $\pm 5\,\text{lb/hr}$, find the least squares estimate of the total flow.

5. The tolerance limits for the measured variables x and y are $1\cdot06 \pm 0\cdot12$ and $2\cdot31 \pm 0\cdot15$ respectively. Find the tolerance limits of the response variable $z = xy$.

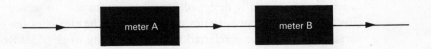

Figure 58

Reference

MANDEL, J. (1964), *The Statistical Analysis of Experimental Data*, Interscience.

Chapter 10
The design and analysis of experiments –
1 Comparative experiments

10.1 **Some basic considerations in experimental design**

The design and analysis of experiments is an extensive subject to which numerous books have been entirely devoted (see references, particularly Davies, 1951, Cochran and Cox, 1957 and Cox, 1958). The problems of design are of course inseparable from those of analysis and it is worth emphasizing from the outset that, unless a sensible design is employed, it may be very difficult or even impossible to obtain valid conclusions from the resulting data. In the next two chapters we will try to illustrate the basic principles of experimental design and try to provide an introduction to the maze of designs which have been proposed. In addition the main methods of analysis will be discussed, including the analysis of variance. Nevertheless the scientist should still be prepared to seek the advice of a statistician if the experiment is at all complicated, not only regarding the analysis of the results but also to select the appropriate design.

Much of the early work on the subject was connected with agricultural experiments, such as comparing the yield from several varieties of wheat. However so many other factors are involved, such as weather conditions, type of soil, position in field, and so on, that unless the experiment is carefully designed, it will be impossible to separate the effects of the different varieties of wheat from the effects of the other variables. The reader is referred to the pioneering work of Sir Ronald Fisher (1951).

In general the techniques described in the next two chapters are useful whenever the effects being investigated are liable to be masked by experimental variation which is outside the control of the scientist. The physicist or chemist working in a laboratory can often keep unwanted variation to a minimum by working with pure materials and by the careful control of variables other than those of direct interest. In this case the methods described here may be of little value. However for large scale industrial experiments complete control is impossible. There is inevitably considerable variation in repeated experiments

because the raw materials vary in quality, and changes occur in the time of day or other environmental conditions. Thus it is highly desirable to use an experimental design which will enable us to separate the effects of interest from the *uncontrolled* or *residual variation*. It is also important to ensure that there is no systematic error in the results. Important techniques for avoiding systematic error and for increasing the precision of the results are:
(a) Randomization,
(b) Replication,
(c) Blocking,
(d) The analysis of covariance.

In Chapter 9 it was stated that the first step in planning an experiment is to formulate a clear statement of the objectives of the test program; the second step is to choose a suitable response variable. We must then make a list of the factors which may affect the value of the response variable. We must also decide how many observations should be taken and what values should be chosen for each factor in each individual test run.

In Chapter 11 we will consider experiments involving several different factors but in this chapter we will concentrate our attention on *comparative experiments* of which the following is an example.

One stage of the turbo-jet cycle consists of burning fuel in the high pressure discharge from the compressor. The efficiency of the burner is of crucial importance and depends, among other things, on the burner discharge temperature. Assume that we have been given the task of designing an experiment to compare two new burners (B_1 and B_2) with a standard burner (B_3).

A similar type of situation would occur if we wanted to compare several different processes for hardening steel. The objective of the experimental programme is to compare the burners in the first situation and to compare the different processes in the second situation. Such experiments are called comparative experiments. Generally we shall refer to the burners, processes or whatever are being compared as *treatments*. In any individual test run only one treatment is present. Note that an individual test run is sometimes called a trial, and that the term experiment is used to denote the whole set of trials and does not refer to an individual test run.

The term *experimental unit* will be used to denote the object on which a particular trial is carried out. In some experiments it is different for each individual test run – this occurs when different drugs are tried on different animals. In these cases we must decide how to assign the different treatments to the different experimental units. In other experiments several tests may be made on the same experimental unit – this occurs when different burners are compared using the same engine. In these cases the individual tests must be carried out sequentially and then it is important to make a careful choice of the order in which the tests should be performed.

Finally, the analysis of the experimental results should give us estimates of the treatment and factor effects and also of the residual variation. In order to carry out such an analysis, it is essential to write down a mathematical model to describe the particular experimental situation. This mathematical model will involve a number of assumptions concerning the relationship between the response variable and the treatments or factors, and, before we go on to consider some experimental designs in greater detail, it will be useful to become familiar with some of the assumptions which are commonly made in formulating mathematical models.

10.2   A mathematical model for simple comparative experiments

We begin by discussing the mathematical model for a simple comparative experiment involving several different treatments. It is usually reasonable to make what is called the assumption of *additivity*. This says that the value of the response variable is equal to

$$
\begin{pmatrix} \text{a quantity depending} \\ \text{on the environmental} \\ \text{conditions and the} \\ \text{experimental unit} \end{pmatrix} + \begin{pmatrix} \text{a quantity depending} \\ \text{on the treatment} \\ \text{used} \end{pmatrix}
$$

If in fact these effects are thought to be multiplicative then we can make a similar assumption by working with the logarithms of the observations.

If there are c treatments, and n observations are made on each treatment, then a suitable mathematical model for this comparative experiment is given by the following:

$$x_{ij} = \mu_i + \varepsilon_{ij} \quad (i = 1, \ldots, c; j = 1, \ldots, n),$$

where

$x_{ij} = j$th observation on treatment i,

$\mu_i = $ average value of observations made on treatment i,

$\varepsilon_{ij} = $ random error.

In addition to the assumption of additivity, it is often necessary to make further assumptions about the distribution of the random errors. Firstly we assume that the errors are normally distributed with mean zero. We further assume that the error variance does not depend on the treatment involved – in other words the error variance is homogeneous. This is usually denoted by σ^2. Finally we assume that successive random errors are uncorrelated.

Similar assumptions to the above will be made in most of the models discussed in the next two chapters. Of course these assumptions will not always be satisfied exactly. However they usually hold sufficiently well to justify the methods of analysis which will be described here. In section 10.9 we shall see how to check that the assumptions are reasonably valid.

10.3 The number of replications

In a comparative experiment the number of observations on each treatment is called the number of replications.

A fundamental principle of such experiments is that it is essential to carry out more than one test on each treatment in order to estimate the size of the experimental error and hence to get some idea of the precision of the estimates of the treatment effects.

This can be illustrated by considering an experiment to compare two methods of increasing the tensile strength of steel beams. Use method A (the first treatment) on one beam and method B (the second treatment) on a similar beam. Suppose the coded results (p.s.i. $\times 10^5$) are the following:

A_1 = strength of beam treated by method A

   = 180,

B_1 = strength of beam treated by method B

   = 168.

Then we would suspect that method A was better than B. But we really have no idea if the difference of 12 between A_1 and B_1 is due to the difference in treatments or is simply due to the natural variability of the strength of steel beams. Even if there is no difference between the two treatments it is highly unlikely that the results will be exactly the same.

Suppose the experiment is repeated and that the following results are obtained:

$A_2 = 176$ $B_2 = 171.$

Then an estimate of the difference between the two treatments is given by

$$\tfrac{1}{2}(A_1 + A_2 - B_1 - B_2) = 8\tfrac{1}{2}. \qquad\qquad\text{10.1}$$

But now we also have two estimates of the residual variation, namely $(A_1 - A_2)$ and $(B_1 - B_2)$. These can be combined in two ways to give

$$\tfrac{1}{2}(A_1 - A_2 + B_1 - B_2) = \tfrac{1}{2}, \qquad\qquad\text{10.2}$$

$$\tfrac{1}{2}(A_1 - A_2 - B_1 + B_2) = 3\tfrac{1}{2}. \qquad\qquad\text{10.3}$$

The treatment comparison, $8\tfrac{1}{2}$, is larger than the other two comparisons, $\tfrac{1}{2}$ and $3\tfrac{1}{2}$, and this is a definite indication that treatment A is better than treatment B. Since there are only two treatments we can compare the treatment effect with the residual variation by means of a two-tailed t-test, provided that it is reasonable to assume that successive observations on a particular treatment are normally distributed. It can be shown that the estimate of the residual standard deviation is given by

$$s = \sqrt{\left[\frac{(A_1 - A_2)^2 + (B_1 - B_2)^2}{4}\right]} = 2\tfrac{1}{2}.$$

The standard error of the estimate of the treatment difference is given by

$$s\sqrt{\left(\frac{1}{\text{number of observations on A}} + \frac{1}{\text{number of observations on B}}\right)}$$

$$= 2\tfrac{1}{2}.$$

Thus the value of the t-test statistic is $8\tfrac{1}{2}/2\tfrac{1}{2} = 3\cdot4$. But the estimate of s is based on just two degrees of freedom (see below) and $t_{0\cdot025,2} = 4\cdot30$ so that the result is not significant at the 5 per cent level. Nevertheless we still suspect that there is a difference between the two methods and so it would be advisable to make some more observations in order to improve the power of the test.

It is difficult to give general realistic advice about the number of replications required in a specific situation. The larger the number of replications, the smaller will be the standard error of the difference between two treatment means $(2\sigma/\sqrt{n})$ and hence the larger the power of the resulting significance test. Clearly if it is desired to detect a 'small' difference between two treatments then a 'large' number of replications will be required, and so it may be possible to determine the required number of replications by considering the power of the test. Some general comments on this problem are given in Cochran and Cox (1957).

In the above example there were just two treatments. More generally, three or more treatments may be compared and then it is necessary to use the technique called the analysis of variance rather than a whole series of t-tests. This technique is described in section 10.7.

At this point it is convenient to re-introduce the concept of degrees of freedom, which, in experimental design, can be thought of as the number of comparisons available. In the example discussed above, there were four measurements A_1, A_2, B_1 and B_2, and these can be combined to give exactly three independent comparisons as given in equations **10.1**, **10.2** and **10.3**. Any other comparison of the four observations, for which the sum of the coefficients is zero, can be obtained from these comparisons. For example $(A_1 - A_2)$ can be obtained by adding **10.2** to **10.3**. The number of independent comparisons is equal to the number of degrees of freedom of the observations; in this case just three. This, of course, is the same as the number of degrees of freedom of the standard deviation of all four observations.

One of the degrees of freedom corresponds to the comparison between the two treatments; namely equation **10.1**. The two remaining degrees of freedom correspond to the residual variation and this is the number of degrees of freedom of the residual standard deviation, s. The two comparisons are $(A_1 - A_2)$ and $(B_1 - B_2)$ or alternatively the equations **10.2** and **10.3**. The over-all degrees of freedom (abbreviated d.f.) can be tabulated as follows.

	d.f.
Treatments	1
Residual	2
Total	3

More generally, if n observations are made on each of c treatments, the total number of degrees of freedom can be broken down as follows.

	d.f.
Treatments	$c-1$
Residual	$c(n-1)$
Total	$cn-1$

229   The number of replications

The notion of degrees of freedom will occur repeatedly throughout the next two chapters, and it is to be hoped that the student is familiar with the idea by now.

10.4 Randomization

This section is concerned with the most important basic principle of good experimentation, namely *randomization*. We have already seen that there is often a substantial amount of uncontrolled or residual variation in any experiment, so no test can ever be repeated exactly. This residual variation is caused by factors which have not been (or cannot be) controlled by the experimenter. These factors are called *uncontrolled* or *nuisance factors*. Some of these nuisance factors are functions of time, place and the experimental units, and they are liable to produce trends in the data quite apart from the effects we are looking for. This would mean that successive error terms would be correlated so that one of the assumptions mentioned in section 10.2 would not apply. Fortunately there is a simple but powerful solution which ensures as far as possible that there is no systematic error in the results. This technique is called randomization and should be used wherever possible.

If the tests are performed sequentially in time, then the order of the tests should be randomized. If a different experimental unit is used in each test, then the unit should be selected randomly. Both these operations can be done by using a table of random numbers.

For example, let us look once again at the experiment for comparing three burners for a turbo-jet engine. If several tests are made on the same engine, the engine may gradually deteriorate resulting in a gradual decrease in performance. If we begin with several tests on B_1, continue with several tests on B_2 and finish with several tests on B_3, burner B_3 will then be at a serious disadvantage compared with B_1. We may mistakenly attribute the decline in performance to the burners used later in the test programme, rather than to the ageing of the engine. It is easy to think of other factors which might produce similar trends in the data. The meters and gauges may develop calibration shifts later in the test programme; the fuel supply may be replenished with fuel with slightly different properties; the engine performance is sensitive to the ambient temperature and pressure and this effect can only be partially removed by a correction factor, so there may be spurious differences between day and night and from hot days to cold days. All these effects may result in a built-in systematic error unless the experiment is randomized. This is done as follows.

Suppose we decide to make six observations on each burner. Assign numbers 1 to 6 to B_1, 7 to 12 to B_2 and 13 to 18 to B_3. Enter a table of random numbers at any point and take numbers from the table by moving in a pre-arranged direction, taking two digits at a time. If the number is between 19 and 99 it is discarded. However if we obtain say 16, then the first test is carried out on B_3, and if the second number is say 04, the second test is carried out on B_1. And so on.

Considering another example, suppose that we wanted to compare three anti-corrosion paints. Then it would be stupid to paint six bridges in England with paint A, six bridges in Wales with paint B and six bridges in Scotland with paint C. If paint A appeared to give the best results we would be unable to tell if this was because paint A was 'best', or simply because the weather in England is better than the weather in Wales or Scotland. Instead the three treatments A, B and C could be randomly assigned to the eighteen experimental units (the bridges), or better still a randomized block experiment could be employed as described in section 10.10.

Of course the use of randomization is sometimes restricted by the experimental situation. It may force additional assembly and dis-assembly of the machinery, and it may consume more time. The engineering statistician must weigh the risk of error in his conclusions against the ease of testing. The risk of error without some form of randomization may be very great unless the experimental procedures are extremely well established from countless experiments.

Randomization may sometimes be impossible. For example, the significantly high correlation between smoking and lung cancer does not necessarily prove a causal relationship between the two. One method of proving or disproving this would be to carry out the following randomized experiment. Select several thousand children and randomly divide them into two groups, one to smoke and one not to smoke. If the smoking group developed a significantly higher incidence of lung cancer over the years, it would prove statistically that there really is a causal relationship. But an experiment of this type is of course completely out of the question. Therefore without randomization we can only say that the case is not proven, but that it would still be sensible to give up smoking.

10.5 **The analysis of a randomized comparative experiment**

Let us suppose that n observations are taken on each of c treatments.

The appropriate mathematical model is given by

$$x_{ij} = \mu_i + \varepsilon_{ij} \qquad (i = 1, \ldots, c; j = 1, \ldots, n).$$

We will call this model 10A. The resulting data can be tabulated as follows.

Table 15

	Observations	Total	Sample mean	Population mean	Sample variance
Treatment 1	$x_{11}, x_{12}, \ldots, x_{1n}$	T_1	$\bar{x}_1$	μ_1	s_1^2
Treatment 2	$x_{21}, x_{22}, \ldots, x_{2n}$	T_2	$\bar{x}_2$	μ_2	s_2^2
$\vdots$	$\vdots$	$\vdots$	$\vdots$	$\vdots$	$\vdots$
Treatment c	$x_{c1}, x_{c2}, \ldots, x_{cn}$	T_c	$\bar{x}_c$	μ_c	s_c^2

We have

$$T_i = \sum_{j=1}^{n} x_{ij},$$

$$\bar{x}_i = \frac{T_i}{n},$$

$$s_i^2 = \sum_{j=1}^{n} \frac{(x_{ij} - \bar{x}_i)^2}{n-1}.$$

It is easy to show that the average observation on treatment i is the best point estimate of μ_i.

Thus $\bar{x}_i = \hat{\mu}_i$.

Our over-all aim is to compare these observed treatment means. If they are 'close together', the differences may be simply due to the residual variation. On the other hand if they are widely separated, then there probably is a significant difference between the treatments. We will describe two methods of data testing used to decide whether all the theoretical treatment means, μ_i, are equal. However it will sometimes be 'obvious' just by looking at the data that one treatment is much better than all the other treatments. If we are simply concerned with finding the 'best' treatment, rather than with finding good estimates of the treatment differences, then there may be no point in carrying out further analysis. Conversely if there are only small differences between the observed treatment means, and if these are judged not to be of practical importance, then there may also be no point in carrying out further analysis. (It is to be hoped that by now the reader appreciates the distinction between statistical and practical significance.)

For many purposes it is more convenient to work with the following model. This is equivalent to model 10A.

$$x_{ij} = \mu + t_i + \varepsilon_{ij} \qquad (i = 1, \ldots, c; j = 1, \ldots, n),$$

where

μ = over-all average,

t_i = effect of ith treatment,

ε_{ij} = random error.

We will call this model 10B. In terms of model 10A we have

$$\mu + t_i = \mu_i.$$

Model 10B appears to involve one more unknown parameter. However, as μ is the over-all average, we have the restriction that $\sum\limits_{i=1}^{c} t_i = 0$, so both models have the same number of independent parameters.

The best point estimates of these new parameters are as follows.

Let $T = \sum\limits_{i=1}^{c} T_i$ = (grand total of the observations)

and $\quad \bar{x} = \dfrac{T}{cn}$ = (over-all observed mean).

Then the best point estimate of μ is given by $\hat{\mu} = \bar{x}$, and the best point estimate of t_i is given by

$$\hat{t}_i = \bar{x}_i - \bar{x}.$$

Example 1

Steel wire was made by four manufacturers A, B, C and D. In order to compare their products ten samples were randomly drawn from a batch of wire made by each manufacturer and the strength of each piece of wire was measured. The (coded) values are given overleaf.

	A	B	C	D
	55	70	70	90
	50	80	60	115
	80	85	65	80
	60	105	75	70
	70	65	90	95
	75	100	40	100
	40	90	95	105
	45	95	70	90
	80	100	65	100
	70	70	75	60
Total	625	860	705	905
Mean	62·5	86·0	70·5	90·5
Variance	212·5	184·0	253·8	274·7

$T = 3095$ $\bar{x} = 77·4$.

In terms of model 10A we will denote the mean strength of wire made by A, B, C and D to be μ_1, μ_2, μ_3 and μ_4 respectively. The estimates of these parameters are given by $\hat{\mu}_1 = 62·5$, $\hat{\mu}_2 = 86·0$, $\hat{\mu}_3 = 70·5$ and $\hat{\mu}_4 = 90·5$.

In terms of the equivalent model 10B we will denote the over-all mean strength by μ and the effects of A, B, C and D by t_1, t_2, t_3 and t_4. The estimates of these parameters are given by

$$\hat{\mu} = 77·4 \qquad \hat{t}_1 = -14·9 \qquad \hat{t}_2 = 8·6 \qquad \hat{t}_3 = -6·9 \qquad \hat{t}_4 = 13·1$$

Note that the sum of these treatment effects is zero – except for a round-off error of 0·1.

We have now completed the initial stages of the analysis which consists of setting up a mathematical model to describe the particular situation and obtaining point estimates of the unknown parameters. The next problem is to decide whether or not there is a significant difference between the observed treatment means. In other words we want to test the hypothesis

$$H_0 : \mu_1 = \mu_2 = \ldots = \mu_c \qquad \text{in terms of model 10A}$$

or

$$H_0 : \text{all } t_i = 0 \qquad \text{in terms of model 10B.}$$

If there are just two treatments then we can use a t-test to compare $\bar{x}_1$ with $\bar{x}_2$. Therefore with more than two treatments it might be supposed that we should perform a whole series of t-tests by testing the difference between each pair of means. But this is not so as the over-all level of significance is affected by the number of tests which are performed. If the significance level is 5 per cent and there are say eight treatments then we have to perform twenty-eight tests ($= {}^8C_2$), and the probability that *at least* one test will give a significant result if H_0 is true is approximately $1-(1-0{\cdot}05)^{28} = 0{\cdot}76$. This result is not unexpected when we consider that, even if H_0 is true, the larger the number of treatments the larger the average difference between the largest and smallest observed treatment means will be. In other words the range of the treatment means increases with the number of treatments.

We will describe two methods of testing H_0. The first method is based on the *range* of the treatment means. The second method is a much more general technique called the *analysis of variance*, which will also be used when testing similar hypotheses in many other types of experimental design.

10.6   The range test

We begin by defining the *studentized range*. Let $x_1, x_2, \ldots, x_c$ be a random sample size c from a normal distribution with variance σ^2. The average value of the sample range will be directly proportional to σ and so the sampling distribution of the range can be obtained by considering

$$\frac{\text{range}(x_1, x_2, \ldots, x_c)}{\sigma}.$$

In most practical problems the parameter σ is unknown. However let us suppose that we have an independent estimate, s^2, of σ^2, which is based on v degrees of freedom. Then the studentized range is given by

$$q = \frac{\text{range}(x_1, x_2, \ldots, x_c)}{s}.$$

The sampling distribution of q has been evaluated for different values of v and c, and tables of the percentage points are given in Appendix B. The point $q_\alpha(c, v)$ is such that

$$\text{probability}(q > q_\alpha) = \alpha.$$

We will now describe how the studentized range is used in a comparative experiment. We have c treatment means each of which is the mean of n observations. The null hypothesis, which we wish to test, is given by

$$H_0 : \text{all } t_i = 0 \qquad \text{in terms of model 10B}.$$

If H_0 is true the observed treatment means will be a random sample of size c from a normal distribution with variance σ^2/n. We can obtain an independent estimate of σ^2 by considering the observed treatment variances. The estimated variance of observations on treatment i is given by

$$s_i^2 = \sum_{j=1}^{n} \frac{(x_{ij} - \bar{x}_i)^2}{n-1}$$

and this is an estimate of σ^2. Thus the combined estimate of σ^2 from the variation within groups is given by

$$s^2 = \sum_{i=1}^{c} \frac{s_i^2}{c} = \frac{\sum_{i=1}^{c} \sum_{j=1}^{n} (x_{ij} - \bar{x}_i)^2}{c(n-1)}.$$

This estimate of σ^2 has $c(n-1)$ degrees of freedom. Note that this estimate only depends on the variation within each group and does not depend on the variation between treatments (between groups). Thus this quantity is an estimate of σ^2 whether or not H_0 is true. An estimate of the standard error of the treatment means is given by $s/\sqrt{n}$ and so the ratio

$$q = \frac{\text{range}(\bar{x}_1, \bar{x}_2, \ldots, \bar{x}_c)}{s/\sqrt{n}}$$

will have the same distribution as the studentized range provided that H_0 is true. However if H_0 is not true we expect the range of the treatment means to be higher than expected giving a 'large' value of q. If the observed value of q, which will be denoted by q_0, is higher than $q_\alpha\{c, c(n-1)\}$, then the result is significant at the α level of significance.

Example 1 continued

We have the following information

	Sample mean	Sample variance
Manufacturer A	62·5	212·5
Manufacturer B	86·0	184·0
Manufacturer C	70·5	253·8
Manufacturer D	90·5	274·7

Largest treatment mean = 90·5, smallest treatment mean = 62·5,

range = $90·5 - 62·5 = 28·0$,

$$s^2 = \frac{212 \cdot 5 + 184 \cdot 0 + 253 \cdot 8 + 274 \cdot 7}{4}$$

$$= 231 \cdot 3.$$

This is based on thirty-six degrees of freedom.

$$q_0 = \frac{28 \cdot 0}{\sqrt{(231 \cdot 3/10)}} = 5 \cdot 83,$$

but $q_{0 \cdot 05}(4, 36) = 3 \cdot 80$.

Thus the result is significant at the 5 per cent level. The data suggests that there really is a significant difference between the strength of wire made by the four manufacturers.

10.7 One-way analysis of variance

This section describes a second method of testing the hypothesis that there is no difference between a number of treatments. The total variation of the observations is partitioned into two components, one measuring the variability between the group means, $\bar{x}_1, \bar{x}_2, \ldots, \bar{x}_c$ and the other measuring the variation within each group. These two components are compared by means of an F-test.

The procedure of comparing different components of variation is called the analysis of variance. In the above situation the observations are divided into c mutually exclusive categories and this is called a one-way classification. We then have a *one-way analysis of variance*. It is a little more complicated than the range test but is often more efficient and has the advantage that a similar technique can be applied to more complex situations where the observations are classified by two or more criteria.

We have already seen that the combined estimate of σ^2 from the variation within groups is given by

$$s^2 = \sum_{i=1}^{c} \frac{s_i^2}{c} = \sum_{i=1}^{c} \sum_{j=1}^{n} \frac{(x_{ij} - \bar{x}_i)^2}{c(n-1)}$$

and this is based on $c(n-1)$ degrees of freedom.

We now look at the variation between groups. The observed variance of the treatment means is given by

$$\sum_{i=1}^{c} \frac{(\bar{x}_i - \bar{x})^2}{c-1}.$$

If the null hypothesis is true, this is an estimate of σ^2/n, since the standard error of the treatments means will be $\sigma/\sqrt{n}$. Thus

$$s_B^2 = n \sum_{i=1}^{c} \frac{(\bar{x}_i - \bar{x})^2}{c-1}$$

is an estimate of σ^2 based on $c-1$ degrees of freedom.

If H_0 is true both s^2 and s_B^2 are estimates of σ^2 and the ratio $F = s_B^2/s^2$ will follow an F-distribution with $c-1$ and $c(n-1)$ degrees of freedom. On the other hand, if H_0 is not true, s^2 will still be an estimate of σ^2 but s_B^2 will be increased by the treatment differences and so the F-ratio may be significantly large. In this case we reject H_0 and conclude that there is evidence of a difference between the treatment effects.

The usual way to obtain the F-ratio is to calculate the following quantities and enter them in what is called an analysis of variance (ANOVA) table.

Table 16
One-way ANOVA

Source of variation	Sum of squares	d.f.	Mean square
Between groups (between treatments)	$n \sum_{i=1}^{c} (\bar{x}_i - \bar{x})^2$	$c-1$	s_B^2
Within groups (residual variation)	$\sum_{i=1}^{c} \sum_{j=1}^{n} (x_{ij} - \bar{x}_i)^2$	$c(n-1)$	s^2
Total variation	$\sum_{i=1}^{c} \sum_{j=1}^{n} (x_{ij} - \bar{x})^2$	$cn-1$	

The two mean squares, s_B^2 and s^2, are obtained by dividing the appropriate sum of squares by the appropriate number of degrees of freedom.

An important feature of the ANOVA table is that the total sum of squares is equal to the sum of the between-group and within-group sum of squares. This can be shown as follows. We have

$$x_{ij} - \bar{x} = (x_{ij} - \bar{x}_i) + (\bar{x}_i - \bar{x}).$$

Squaring both sides and summing over all values of i and j, we find

$$\sum_{i,j} (x_{ij} - \bar{x})^2 = \sum_{i,j} (x_{ij} - \bar{x}_i)^2 + \sum_{i,j} (\bar{x}_i - \bar{x})^2 + 2\sum_{i,j} (x_{ij} - \bar{x}_i)(\bar{x}_i - \bar{x}).$$

However the sum of the cross product terms is zero since

$$\sum_{i=1}^{c} \sum_{j=1}^{n} (x_{ij} - \bar{x}_i)(\bar{x}_i - \bar{x}) = \sum_{i=1}^{c} \left\{ (\bar{x}_i - \bar{x}) \sum_{j=1}^{n} (x_{ij} - \bar{x}_i) \right\}$$

and $\displaystyle\sum_{j=1}^{n} (x_{ij} - \bar{x}_i) = 0$ for each i.

Also $\displaystyle\sum_{i=1}^{c} \sum_{j=1}^{n} (\bar{x}_i - \bar{x})^2 = n \sum_{i=1}^{c} (\bar{x}_i - \bar{x})^2$

and we have the required result.

Another important feature of the ANOVA table is that the total number of degrees of freedom is equal to the sum of the between group and within group degrees of freedom. The actual computation proceeds as follows.

Firstly calculate the group totals $T_i(i = 1$ to $c)$ and the over-all total T. Hence find the group means $\bar{x}_i(i = 1$ to $c)$ and the over-all mean $\bar{x}$. Also calculate $\sum_{i,j} x_{ij}^2$, which is sometimes called the total uncorrected sum of squares. Secondly calculate the total sum of squares (sometimes called the total corrected sum of squares) which is given by

$$\sum_{i,j} (x_{ij} - \bar{x})^2 = \sum_{i,j} x_{ij}^2 - cn\bar{x}^2$$

$$= \sum_{i,j} x_{ij}^2 - \frac{T^2}{cn}.$$

The quantity $cn\bar{x}^2$ or T^2/cn is sometimes called the correction factor. Thirdly calculate the between-group sum of squares which is given by

$$n \sum_{i=1}^{c} (\bar{x}_i - \bar{x})^2 = n \sum_{i=1}^{c} \bar{x}_i^2 - cn\bar{x}^2$$

$$= \frac{1}{n} \sum_{i=1}^{c} T_i^2 - \frac{T^2}{cn}.$$

Note that the same correction factor is present. Lastly the residual sum of squares can be obtained by subtraction.

Example 1 continued

We have $\bar{x}_1 = 62.5$, $\bar{x}_2 = 86.0$, $\bar{x}_3 = 70.5$, $\bar{x}_4 = 90.5$, $\bar{x} = 77.4$. Also $\sum_{i,j} x_{ij}^2 = 252975$ $n \sum_{i=1}^{c} \bar{x}_i^2 = 244627$.

and

$$cn\bar{x}^2 = 40 \times 77.4 \times 77.4 = 239476.$$

$$\text{(The total corrected sum of squares)} = 252975 - 239476$$

$$= 13499.$$

$$\text{(The between-group sum of squares)} = 244627 - 239476$$

$$= 5151,$$

$$\text{(the residual sum of squares)} = 13499 - 5151$$

$$= 8348.$$

We can now construct the ANOVA table.

Source	Sum of squares	d.f.	Mean square
Treatments	5151	3	1717
Residual	8348	36	231·3
Total	13499	39	

$$F\text{-ratio} = \frac{1717}{231.3}$$

$$= 7.41.$$

$$F_{0.01,3,36} = 4.39.$$

Thus the result is significant at the 1 per cent level and we reject the hypothesis that there is no difference between wire made by the four manufacturers. This, of course, is the same conclusion that was obtained with a range test. The reader will also notice that the residual mean square is the same as the estimate of σ^2 which was obtained in the range test. However the way it is computed is completely different.

It is useful to know the average or expected values of the treatment and residual mean squares. Since s^2 is an unbiased estimate of σ^2 we have

$E(\text{residual mean square}) = E(s^2) = \sigma^2,$

whether or not H_0 is true. It can also be shown (see Exercise 1) that

$E(\text{treatment mean square}) = E(s_B^2)$

$$= \sigma^2 + n \sum_{i=1}^{c} \frac{t_i^2}{c-1}$$

Thus $E(s_B^2) = \sigma^2$ only if H_0 is true. If H_0 is not true, then the larger the treatment effects are, the larger we can expect the treatment mean square and hence the F-ratio to be.

10.8 Follow-up study of the treatment means

The next problem is to decide what to do when a one-way ANOVA (or range test) indicates that there is a significant difference between the treatments. Some typical questions which we might ask are the following.

(1) Is one treatment much better than all the other treatments?

(2) Is one treatment much worse than all the other treatments?

(3) Can the treatments be divided into several homogeneous groups in each of which the treatments are not significantly different?

One way of answering these questions is to calculate a quantity called the *least significant difference*. The variance of the difference between two treatment means is given by

$$\text{var}(\bar{x}_i - \bar{x}_j) = \sigma^2 \left(\frac{1}{n} + \frac{1}{n} \right).$$

Thus an estimate of the standard error of the difference between the two treatments is given by

$s \sqrt{\left(\dfrac{2}{n}\right)}$ where s is the square root of the residual mean square. Thus the least difference between two means which is significant at the α level is

$$s \sqrt{\left(\frac{2}{n}\right)} \, t_{\frac{1}{2}\alpha, c(n-1)},$$

since the estimate s of σ is based on $c(n-1)$ d.f. The treatments means are arranged in order of magnitude and the difference between any pair of means can be compared with the least significant difference. If the gap between any two means is less than the least significant difference then the treatments are not significantly different. However if the gap between two successive means is larger than the least significant difference then we have a division between two groups of treatments. One way of recording the results is to draw lines underneath the means so that any two means which have the same underlining are not significantly different.

Example 1 continued

The four means are

$$\bar{x}_A = 62.5 \qquad \bar{x}_B = 86.0$$

$$\bar{x}_C = 70.5 \qquad \bar{x}_D = 90.5.$$

The estimate s of σ is given by

$$s = \sqrt{231.3} = 15.2$$

and is based on thirty-six degrees of freedom.

Choose a significance level of 5 per cent. Then the least significant difference is given by

$$t_{0.025}s\sqrt{\tfrac{2}{10}} = 13.8.$$

Arranging the treatment means in order of magnitude we have

$$\bar{x}_A < \bar{x}_C < \bar{x}_B < \bar{x}_D,$$

but $(\bar{x}_D - \bar{x}_B)$ and $(\bar{x}_C - \bar{x}_A)$ are both less than 13.8, whereas $(\bar{x}_B - \bar{x}_C)$ is greater than 13.8. Thus we conclude that manufacturers B and D are better than A and C but that there is not a significant difference between B and D. We can express these results as follows.

$$\underline{\bar{x}_A \quad \bar{x}_C} \qquad \underline{\bar{x}_B \quad \bar{x}_D}$$

The astute reader will notice that the above procedure is nothing more than a multiple t-test. Earlier in the chapter, we decided that this could not be used to test the hypothesis that the treatment means are not significantly different. However it is often used as a follow-up technique to see how the treatments are grouped, once the null hypothesis has been rejected by an analysis of variance or range test. The method will still tend to give too many significant differences, but it is simple to use and for this reason is often preferred to such methods as Duncan's multiple range test, (see, for example, Wine, 1964), which give more reliable results.

If so desired we can also use the t-distribution to calculate confidence intervals for the true treatment means. The $100(1 - \alpha)$ per cent confidence interval for μ_i is given by $\bar{x}_i \pm t_{\alpha/2, c(n-1)} s/\sqrt{n}$.

Example 1 continued

$$t_{0.025, 36} = 2{\cdot}03 \qquad s = 15{\cdot}2.$$

The 95 per cent confidence intervals for μ_A, μ_B, μ_C and μ_D are given by

$$62{\cdot}5 \pm 9{\cdot}8 \qquad 86{\cdot}0 \pm 9{\cdot}8$$

$$70{\cdot}5 \pm 9{\cdot}8 \qquad 90{\cdot}5 \pm 9{\cdot}8.$$

In the above statements the risk that a particular μ_i will be outside its confidence interval is α. But we have formed c such intervals. Thus the expected number of mistakes will be $c\alpha$. In other words the risk that *at least* one of the above statements is false will be much higher than α and is in fact close to $1 - (1 - \alpha)^c$. In order to form confidence intervals so that the overall risk of making one or more incorrect statements is say α', we must calculate the corresponding value of α from the relation

$$\alpha' = 1 - (1 - \alpha)^c.$$

10.9 Verifying the model

A number of assumptions have been incorporated in the mathematical model 10A or 10B. If any of these assumptions are false, some or all of the preceding analysis may be invalidated. Some of the assumptions can be checked by looking at the *residuals*. In a simple comparative experiment the residual variation is the variation not accounted for by the treatment effects. Thus the residual of a particular observation is the difference between the observation and the average observation on that particular treatment.

$$r_{ij} = x_{ij} - \bar{x}_i \qquad (i = 1, \ldots, c; j = 1, \ldots, n).$$

One of the worst possibilities is that the residuals may not be random. For example, we may get a series of negative residuals followed by a series of positive residuals. This would mean that an uncontrolled factor was systematically affecting the results which would make any conclusions suspect. Fortunately the possibility of non-randomness can often be overcome by randomization. Alternatively it can be overcome by a technique called blocking which will be described in the next section.

Another possibility is that the errors may not be normally distributed. Fortunately the F-test and the t-test are both robust to departures from the normality assumption. However it will occasionally be necessary to transform the data in order to avoid a distribution of residuals which is markedly skew. The logarithmic transformation is commonly used.

A further possibility is that the error variance does vary from treatment to treatment. In this case the treatment means should be compared not with a combined estimate of the variance but with the individual variance estimates of the particular treatments. Alternatively it may be possible to transform the data so that the error variance is constant. In particular if we suspect that the treatment standard deviation is directly proportional to the treatment mean then we can make the variance homogeneous with a logarithmic transformation. Visual inspection of the group variances is usually sufficient and in Example 1 there is no evidence that the variance is not homogeneous. More generally the treatment variances $s_1^2, s_2^2, \ldots, s_c^2$ are estimates of $\sigma_1^2, \sigma_2^2, \ldots, \sigma_c^2$. The standard method of testing the hypothesis $\sigma_1^2 = \sigma_2^2 = \ldots = \sigma_c^2$ is by Bartlett's test (Davies, 1951).

10.10 The randomized block experiment

We have seen that randomization is a simple but effective way of eliminating systematic error in many experiments. But the experimenter may still be aware of one particular environmental factor which contributes substantially to the uncontrolled variation. For example, if the experiment extends over several days, then observations made on the same day will show better agreement than those made on different days. Thus there is a danger of introducing systematic error unless the randomization procedure happens to give an equal number of tests on each treatment in each day. We can avoid this danger by dividing the tests into groups which are 'close together' in some way; for example, measurements made by the same operator or measurements made on the same batch of raw material. These groups are called *blocks*. If an equal number of measurements is made on each treatment in each block and if the order of tests within a block is randomized, then the experiment is called a *randomized block experiment*.

The technique of blocking is a very useful method of increasing the precision of comparative experiments. All comparisons are made within a block and not from block to block. For example, let us look once again at the experiment for comparing the three anti-corrosion paints. A fully randomized design might happen to assign the treatments in the following way.

England: 4 tests on A   1 test on B   1 test on C,
Wales: 1 test on A   3 tests on B   2 tests on C,
Scotland: 1 test on A   2 tests on B   3 tests on C.

Thus if the weather is indeed better in England, then it would again be true that paint A has an unfair advantage over the other two paints. The 'obvious' way to carry out the experiment is to make two observations on each paint in England, Scotland and Wales. Within each country the paints will be assigned randomly to the six bridges. Here the countries are the blocks and we can now compare the observations within a particular country.

Let us also reconsider the experiment in which three burners are compared in a turbo-jet engine. Suppose that four observations are to be made on each burner, but that only six observations can be made in one day. If we suspect a significant difference from day to day, we must try to design the experiment so that an equal number of observations are made on each burner in each day. In other words on day 1 two observations should be made on each of B_1, B_2 and B_3, and similarly on day 2. Within each day the order of the six experiments is randomized and we then have a randomized block experiment. A similar situation would arise if the tests were to be carried out on two different engines. Since there is likely to be a difference between the two engines, an equal number of observations must be made on each burner in each engine.

It is easy to think of many other experimental situations in which blocking is necessary and the randomized block experiment is probably the most important type of comparative experiment. If there is just one observation on each of c treatments in each of r blocks a suitable mathematical model is given by the following:

$$x_{ij} = \mu + b_i + t_j + \varepsilon_{ij} \qquad (i = 1, \ldots, r; j = 1, \ldots, c),$$

where

x_{ij} = observation on treatment j in block i,

μ = over-all average of the response variable,

b_i = effect of the ith block,

t_j = effect of the jth treatment,

ε_{ij} = random error.

Since μ is the over-all mean, the treatment and block effects must be such that

$$\sum_{i=1}^{r} b_i = \sum_{j=1}^{c} t_j = 0.$$

Notice that the model assumes that the treatment and block effects are additive. It is also convenient to assume that the errors are normally distributed with mean zero and constant variance σ^2 and that the errors are independent.

The data can be tabulated as follows.

Table 17

	Treatment 1	...	Treatment c	Row total	Row average
Block 1	x_{11}	...	x_{1c}	$T_1.$	$\bar{x}_1.$
Block 2	x_{21}	...	x_{2c}	$T_2.$	$\bar{x}_2.$
	.	.	.	.	.
Block r	x_{r1}	...	x_{rc}	$T_r.$	$\bar{x}_r.$
Column total	$T_{.1}$	...	$T_{.c}$		
Column average	$\bar{x}_{.1}$	...	$\bar{x}_{.c}$		

Grand total of the observations = $T = \sum_i T_i. = \sum_j T_{.j}$.

(It is a good idea to check that the sum of the row totals is equal to the sum of the column totals.)

(Over-all average) = $\bar{x} = \dfrac{T}{rc}$.

The best point estimates of the unknown parameters are

$\hat{\mu} = \bar{x}$,

$\hat{b}_i = \bar{x}_i. - \bar{x} \quad (i = 1 \text{ to } r)$,

$\hat{t}_j = \bar{x}_{.j} - \bar{x} \quad (j = 1 \text{ to } c)$.

Example 2

In order to compare three burners, B_1, B_2 and B_3, one observation is made on each burner on each of four successive days. The data is tabulated below.

	B_1	B_2	B_3	Row total	Row average
Day 1 (Block 1)	21	23	24	68	22·67
Day 2 (Block 2)	18	17	23	58	19·33
Day 3 (Block 3)	18	21	20	59	19·67
Day 4 (Block 4)	17	20	22	59	19·67
Column total	74	81	89		
Column average	18·50	20·25	22·25		

$T = 244$ $\bar{x} = 20 \cdot 33 = \hat{\mu}$.

$\hat{t}_1 = (\text{effect of } B_1) = 18 \cdot 50 - 20 \cdot 33 = -1 \cdot 83$,

$\hat{t}_2 = 20 \cdot 25 - 20 \cdot 33 = -0 \cdot 08$,

$\hat{t}_3 = 22 \cdot 25 - 20 \cdot 33 = 1 \cdot 92$,

$\hat{b}_1 = (\text{effect of day 1}) = 22 \cdot 67 - 20 \cdot 33 = 2 \cdot 34$,

$\hat{b}_2 = 19 \cdot 33 - 20 \cdot 33 = -1 \cdot 00$,

$\hat{b}_3 = 19 \cdot 67 - 20 \cdot 33 = -0 \cdot 66$,

$\hat{b}_4 = 19 \cdot 67 - 20 \cdot 33 = -0 \cdot 66$.

Notice that the sum of the treatment effects and the sum of the block effects are both zero, except for round-off errors of $0 \cdot 01$ and $0 \cdot 02$ respectively.

It looks as though burner B_3 gives the best results but this is by no means certain from a visual inspection of the data since results from B_3 are not always larger than those of B_2, even within a particular block. In the next section we shall see how to carry out a two-way analysis of variance in order to see if the differences between the burners are significantly large.

It is important to realise that the estimates of the treatment effects do not depend on the variation between blocks. In order to convince himself of this fact the reader should try adding a constant to all the observations in say block 1 of Example 2. This will alter the block effects and the over-all mean but the estimates of the treatment effects will not change.

10.11 Two-way analysis of variance

In a randomized block experiment the data is classified according to two characteristics in a two-way table. With one observation on each treatment in each block, we have proposed the mathematical model

$$x_{ij} = \mu + b_i + t_j + \varepsilon_{ij} \quad (i = 1 \text{ to } r; j = 1 \text{ to } c)$$

and have seen how to obtain point estimates of the model parameters. The next problem is to see if there is a significant difference between the observed treatment effects. In other words we want to test the hypothesis

$$H_0: \text{all } t_j = 0.$$

The total corrected sum of squares is given by $\sum_{i,j}(x_{ij} - \bar{x})^2$ and this can be split up into three components in a somewhat similar way to that described in the one-way analysis of variance. One of these components measures the variation between treatments, one measures the variation between blocks and the third measures the residual variation. The treatment variation is then compared with the residual variation by means of an F-test. Since the data is classified according to two characteristics (treatment and block), this procedure is called a *two-way analysis of variance*.

The required algebraic identity is

$$\sum_{i,j}(x_{ij} - \bar{x})^2 = c \sum_{i=1}^{r}(\bar{x}_{i.} - \bar{x})^2 + r \sum_{j=1}^{c}(\bar{x}_{.j} - \bar{x})^2 + \sum_{i,j}(x_{ij} - \bar{x}_{i.} - \bar{x}_{.j} + \bar{x})^2$$

and the student is asked to verify this equation in Exercise 3. The first component on the right-hand side of this equation measures the block variation, the second component measures the treatment variation and the third component measures the residual variation.

The corresponding degrees of freedom are as follows.

	d.f.
Blocks (rows)	$r-1$
Treatments (columns)	$c-1$
Residual	$(r-1)(c-1)$
Total	$rc-1$

Each sum of squares is now divided by the appropriate number of degrees of freedom to give the corresponding mean square. It can be shown that the average values of these mean squares are as follows.

$$E(\text{block mean square}) = \sigma^2 + \frac{c}{r-1} \sum_{i=1}^{r} b_i^2,$$

$$E(\text{treatment mean square}) = \sigma^2 + \frac{r}{c-1} \sum_{j=1}^{c} t_j^2,$$

$$E(\text{residual mean square}) = \sigma^2.$$

Neither the treatment mean square nor the residual mean square is affected by whether or not there is a significant variation between blocks. Thus the ratio

$$F = \frac{\text{treatment mean square}}{\text{residual mean square}}$$

will follow an F-distribution with $(c-1)$ and $(r-1)(c-1)$ degrees of freedom if the null hypothesis is true. The observed F-ratio can then be compared with the chosen upper percentage point of this F-distribution. The important point to grasp is that, if there is a block effect, the residual mean square will certainly be smaller in a randomized block experiment than it would have been in a simple randomized experiment. Thus the blocking technique enables us to carry out an F-test which is more sensitive to treatment effects.

The ANOVA table is as follows.

Table 18
Two-way ANOVA

Source of variation	Sum of squares	d.f.	E(mean square)
Treatments (columns)	$r \sum_{j=1}^{c} (\bar{x}_{.j} - \bar{x})^2$	$c-1$	$\sigma^2 + \frac{r}{c-1} \sum t_j^2$
Blocks (rows)	$c \sum_{i=1}^{r} (\bar{x}_{i.} - \bar{x})^2$	$r-1$	$\sigma^2 + \frac{c}{r-1} \sum b_i^2$
Residual	$\sum_{i,j} (x_{ij} - \bar{x}_{i.} - \bar{x}_{.j} + \bar{x})^2$	$(r-1)(c-1)$	σ^2
Total	$\sum_{i,j} (x_{ij} - \bar{x})^2$	$rc-1$	

The actual computation proceeds as follows. First calculate the row totals, the column totals and the over-all total. Hence find the row and column averages and the over-all average; also calculate $\sum x_{ij}^2$ and the correction factor T^2/cr or $cr\bar{x}^2$. Then the total corrected sum of squares is given by

$$\sum (x_{ij} - \bar{x})^2 = \sum x_{ij}^2 - \frac{T^2}{cr}.$$

Secondly the treatment (column) sum of squares is given by

$$r \sum_{j=1}^{c} (\bar{x}_{.j} - \bar{x})^2 = r \sum_{j=1}^{c} \bar{x}_{.j}^2 - \frac{T^2}{cr}$$

$$= \frac{1}{r} \sum_{j=1}^{c} T_{.j}^2 - \frac{T^2}{cr}.$$

Thirdly the block (row) sum of squares is given by

$$c \sum_{i=1}^{r} (\bar{x}_{i.} - \bar{x})^2 = c \sum_{i=1}^{r} \bar{x}_{i.}^2 - \frac{T^2}{cr}$$

$$= \frac{1}{c} \sum_{i=1}^{r} T_{i.}^2 - \frac{T^2}{cr}.$$

Notice that the correction factor appears in all three equations. The residual sum of squares can now be obtained by subtraction.

Each sum of squares is now divided by the appropriate number of degrees of freedom to give the required mean squares. Two F-ratios can now be calculated by dividing first the treatment mean square, and secondly the block mean square, by the residual mean square. The observed treatment F-ratio can then be compared with the upper percentage point of the appropriate F-distribution in order to test the hypothesis that all the treatment effects are zero. If the block effect is of interest we can also test the hypothesis that all the block effects are zero by considering the observed block F-ratio.

Example 2 continued

$$\sum x_{ij}^2 = 5026 \qquad \frac{1}{c} \sum T_{i.}^2 = 4983 \cdot 33 \qquad \frac{1}{r} \sum T_{.j}^2 = 4989 \cdot 50,$$

Correction factor = $4961 \cdot 33$.

Source	Sum of squares	d.f.	Mean square	F-ratio
Treatments	28·17	2	14·08	5·83
Blocks	22·00	3	7·33	3·03
Residual	14·50	6	2·42	
Total	64·67	11		

But $F_{0·05,2,6} = 5·14$. Thus the treatments effects are significant at the 5 per cent level and so we have reasonable evidence that there is a difference between the burners.

We also have $F_{0·05,3,6} = 4·76$ so the block effect is not significant at the 5 per cent level. Thus we have no real evidence that there is a difference between days. Nevertheless the block mean square is three times as large as the residual mean square and so we have increased the precision of the experiment by blocking.

In general, if n observations are made on each treatment in each block a suitable mathematical model is given by

$$x_{ijk} = \mu + b_i + t_j + \varepsilon_{ijk}$$

where $i = 1$ to r, $j = 1$ to c, $k = 1$ to n and $x_{ijk} = k$th observation on the jth treatment in block i. The analysis begins as before by finding the average observation on each treatment, the average observation in each block and the over-all average. This gives estimates of the treatment and block effects as found before. The ANOVA table is shown below.

Table 19

Source	Sum of squares	d.f.	E(mean square)
Treatments	$nr \sum_j (\bar{x}_{.j} - \bar{x})^2$	$c - 1$	$\sigma^2 + nr \sum_j \dfrac{t_j^2}{c-1}$
Blocks	$nc \sum_i (\bar{x}_{i.} - \bar{x})^2$	$r - 1$	$\sigma^2 + nc \sum_i \dfrac{b_i^2}{r-1}$
Residual	by subtraction	by subtraction	σ^2
Total	$\sum_{i,j,k} (x_{ijk} - \bar{x})^2$	$nrc - 1$	

An F-test can now be carried out to test the hypothesis that the treatment effects are zero. Finally the assumptions contained in the model proposed for the randomized block experiment can be checked in a similar way to that described in section 10.9. A full discussion of the examination of residuals is given in section 11.9.

10.12 Latin squares

The randomized block experiment is by far the most important type of comparative experiment. However we will briefly mention two types of modification; namely Latin square designs and balanced incomplete block designs.

Sometimes it is possible to think of two different ways of dividing the experiment into blocks. For example if we decide to use three engines to compare three burners B_1, B_2 and B_3 and to make three tests on each burner then a possible design is the following.

	Time of day		
	Time 1	Time 2	Time 3
Engine 1	B_1	B_2	B_3
Engine 2	B_3	B_1	B_2
Engine 3	B_2	B_3	B_1

The engines and times of day form the two types of block. The significant point is that each burner only appears once in each row and in each column. A design in which each treatment appears exactly once in each row and each column is called a *Latin square design*.

It is easy to construct a Latin square. For example, the following is a 4×4 Latin square.

A B C D

D A B C

C D A B

B C D A

The experiment can be randomized by randomly ordering the rows and columns. The disadvantage of Latin square designs is that the number of both types of block must equal the number of treatments, a condition which is rather restrictive.

The analysis of results from a Latin square design is an extension of the two-way analysis of variance. In addition to calculating the row and column totals it is also necessary to calculate the sum of observations for each treatment.

$T_{(i)}$ = sum of observations on treatment i.

The numbers of rows, columns and treatments are all equal, and will be denoted by m. The row, column and total sums of squares are calculated, as in Table 18, by substituting m for r and c. However we can now extract one more component from the total sum of squares to measure the treatment variation. The treatment sum of squares is given by

$$\frac{1}{m} \sum_{i=1}^{m} T_{(i)}^2 - \frac{T^2}{m^2}.$$

The residual sum of squares can then be obtained by subtraction. The degrees of freedom are as follows.

	$d.f.$
Treatments	$m-1$
Rows	$m-1$
Columns	$m-1$
Residual	m^2-3m+2
Total	m^2-1

The treatment mean square can now be compared with the residual mean square by means of an F-test.

10.13 Balanced incomplete block designs

Sometimes the number of tests which form a homogeneous group is smaller than the number of treatments. Then it is not possible to test each treatment in each block and we have what is called an *incomplete block design.*

For example, suppose that we want to compare three burners in the same engine but that it is only possible to make two test runs in one day. Then the following is a possible design.

Day 1: B_1 and B_2,
Day 2: B_1 and B_3,
Day 3: B_2 and B_3,
Day 4: B_1 and B_2,
Day 5: B_1 and B_3,
Day 6: B_2 and B_3.

The experiment is randomized by tossing a coin to decide which treatment comes first in each block.

An incomplete block design is called *balanced* if each pair of treatments occur together the same number of times. By inspection the above design is seen to be balanced as each burner occurs twice with every other burner. Two treatments can be compared by considering the differences within each block. To illustrate, we will consider the above problem in which there are just two observations in each block.

Let d_1 = difference between the observations on B_1 and B_2 on day 1; similarly for $d_2, d_3, \ldots, d_6$.

Let b_i denote the effect of the ith burner. Then d_1, d_4 are both estimates of $(b_1 - b_2)$; d_2, d_5 are both estimates of $(b_1 - b_3)$; and d_3, d_6 are both estimates of $(b_2 - b_3)$.

But we can obtain better estimates of these differences by combining all the observations. For example, to estimate $(b_1 - b_2)$ we note that

$$d_1 + d_4 + d_2 + d_5$$

is an estimate of $\quad 4b_1 - 2b_2 - 2b_3$,

or $\quad 6b_1 - (2b_1 + 2b_2 + 2b_3)$

and that $\quad -d_1 - d_4 + d_3 + d_6$

is an estimate of $\quad 4b_2 - 2b_1 - 2b_3$,

or $\quad 6b_2 - (2b_1 + 2b_2 + 2b_3)$.

By subtraction we have:

$$2d_1 + 2d_4 + d_2 + d_5 - d_3 - d_6$$

is an estimate of $6(b_1 - b_2)$. Thus the estimate of the difference between burners B_1 and B_2 is given by

$$\tfrac{1}{6}(2d_1 + 2d_4 + d_2 + d_5 - d_3 - d_6).$$

The general estimation of treatment effects and the analysis of variance for a balanced incomplete block design is considered, for example, in Cochran and Cox (1957).

Exercises

1. In a simple comparative experiment, show that the expected value of $n \sum_{i=1}^{c} (\bar{x}_i - \bar{x})^2$ is equal to $(c-1)\sigma^2 + n \sum_{i=1}^{c} t_i^2$. (Hint: $\text{var}(\bar{x}) = \sigma^2/nc$.

$\text{var}(\bar{x}_i) = \sigma^2/n$. Expand $\sum\limits_{i=1}^{c} (\bar{x}_i - \bar{x})^2$ and show $\sum\limits_{i=1}^{c} \bar{x}_i \bar{x} = c\bar{x}^2$). Hence show that the expected value of the treatment mean square is given by $\sigma^2 + n \sum\limits_{i=1}^{c} t_i^2/(c-1)$.

2. If the number of observations on each treatment in a one-way analysis of variance is unequal and there are n_i observations on treatment i, show that

$$\sum_{i=1}^{c} \sum_{j=1}^{n_i} (x_{ij} - \bar{x})^2 = \sum_{i=1}^{c} \sum_{j=1}^{n_i} (x_{ij} - \bar{x}_i)^2 + \sum_{i=1}^{c} n_i (\bar{x}_i - \bar{x})^2.$$

Also verify that the number of degrees of freedom corresponding to the total sum of squares, the treatment sum of squares and the residual sum of squares are $N-1$, $c-1$ and $N-c$ respectively where

$$N = \sum_{i=1}^{c} n_i.$$

3. In a two-way analysis of variance show that

$$\sum_{i=1}^{r} \sum_{j=1}^{c} (x_{ij} - \bar{x})^2 = c \sum_{i=1}^{r} (\bar{x}_{i.} - \bar{x})^2 + r \sum_{j=1}^{c} (\bar{x}_{.j} - \bar{x})^2$$

$$+ \sum_{i=1}^{r} \sum_{j=1}^{c} (x_{ij} - \bar{x}_{i.} - \bar{x}_{.j} + \bar{x})^2.$$

4. Test the hypothesis that the following four sets of data are homogeneous.

Set 1: 7, 8, 7, 10.
Set 2: 6, 8, 9, 7.
Set 3: 11, 10, 12, 9.
Set 4: 12, 10, 9, 10.

5. The following data are the results from a randomized block experiment, which was set up to compare four treatments.

		Treatment			
		1	2	3	4
	1	6	8	7	11
Block	2	5	6	6	8
	3	9	9	11	12

Test the hypothesis that there is no difference between the treatments.

6. The following data resulted from an experiment to compare three burners, B_1, B_2 and B_3. A Latin square design was used as the tests were made on three engines and were spread over three days.

	Engine 1		Engine 2		Engine 3	
Day 1	B_1	16	B_2	17	B_3	20
Day 2	B_2	16	B_3	21	B_1	15
Day 3	B_3	15	B_1	12	B_2	13

Test the hypothesis that there is no difference between the burners.

7. Set up a balanced incomplete block design to compare five treatments in blocks of size three.

References

COCHRAN, W. G., and COX, G. M. (1957), *Experimental Designs*, Wiley, 2nd edn.
COX, D. R. (1958), *Planning of Experiments*, Wiley.
DAVIES, O. L. (ed.) (1951), *Design and Analysis of Industrial Experiments*, Oliver & Boyd, 6th edn.
FISHER, R. A. (1951), *Design of Experiments*, Oliver & Boyd, 6th edn.
WINE, R. L. (1964), *Statistics for Scientists and Engineers*, Prentice-Hall.

Chapter 11
The design and analysis of experiments –
2 Factorial experiments

11.1 Introduction

We now turn our attention to experiments where the response variable depends on several different factors. For example, the yield of a chemical reaction may be affected by changes in the temperature, the pressure and the concentration of one of the chemicals. It may also be affected by whether or not the mixture is agitated during the reaction and by the type of catalyst employed.

A *factor* is any feature of the experimental conditions which is of interest to the experimenter. Factors are of two types. Firstly a *quantitative* factor is one where possible values can be arranged in order of magnitude. Any continuous variable, such as temperature or pressure, is a quantitative factor. Conversely a *qualitative* factor is one whose possible values cannot be arranged in order of magnitude. In the above example the type of catalyst employed is a qualitative factor.

The value that a factor takes in a particular test is called the *level* of the factor. Except for nuisance factors, we will assume that the levels of the factors of interest can be determined by the experimenter. By analogy with comparative experiments, a specific combination of factor levels is called a treatment or *treatment combination*.

It is clear that comparative experiments can be thought of as a special case of factorial experiments. In a simple comparative experiment the individual treatments are the different levels of the one and only factor. A randomized block comparative experiment can also be thought of as a factorial experiment in which two factors are involved. The individual treatments form one factor and the blocks form a second factor.

The over-all objective of a factorial experiment may be to get a general picture of how the response variable is affected by changes in the different factors or to find the over-all combination of factor levels which gives a maximum (or minimum) value of the response variable. The latter problem is discussed in section 11.10.

A common type of test programme is the classical *one-at-a-time experiment* in which the value of the response variable is found for a particular treatment combination, after which the factors are altered one-at-a-time while keeping the other factors at their initial values. Such experiments suffer from several serious defects and we shall see that it is often preferable to run what is called a *complete factorial experiment*, in which one or more trials is made at every possible combination of factor levels. Such an experiment is often simply called a factorial experiment.

11.2 The advantages of complete factorial experiments

The most serious defect of one-at-a-time experiments is that they are unable to detect interactions between factors, whereas complete factorial experiments can. For example, let us suppose that we are interested in finding the yield of a chemical reaction at two temperatures, T_0 and T_1, and at two pressures P_0 and P_1. In a one-at-a-time experiment we would begin by taking one or more observations at $(T_0 P_0)$, $(T_0 P_1)$ and $(T_1 P_0)$. Suppose the average observation is 100 at $(T_0 \ P_0)$, 108 at $(T_0 \ P_1)$ and 105 at $(T_1 \ P_0)$. The question then arises as to what value the response variable will take at $(T_1 P_1)$.

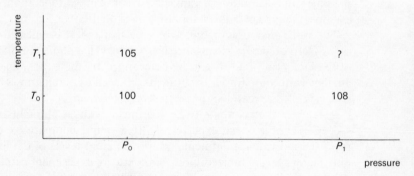

Figure 59   One-at-a-time experiment

If the average observation at $(T_1 P_1)$ happens to be 113, then the effects of the two factors are said to be additive. In other words the difference between the average observation at T_1 and at T_0 is the same for both values of P (and vice versa). But if the average observation at $(T_1 P_1)$ is anything but 113, then the two factors are said to *interact*.

Usually the experimenter will not know if there is an interaction between the two factors and then it is necessary to take measurements at $(T_1 P_1)$ in order to find this out. When this is done we have a complete factorial experiment. Two possible sets of results are shown in Figure 60. If there is no interaction between the two factors then the response curves at different levels of the pressure will be parallel.

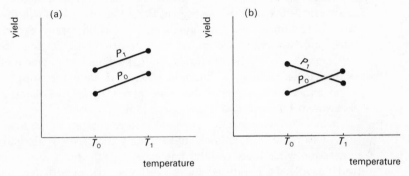

Figure 60   Two possible sets of results
(a) lines parallel, no interaction (b) lines not parallel, interaction

The results from a complete factorial experiment can be combined to estimate the effects of the individual factors, called the *main effects*, and also to estimate the interactions between the factors. For example, let us suppose the average observation at $T_1 P_1$ turns out to be 109. The main effect of pressure is defined to be the average difference between the yield at P_1 and at P_0 which is $108.5 - 102.5 = 6.0$. Similarly the main effect of temperature is given by $107 - 104 = 3$. The interaction between the two factors is defined to be half the difference between the temperature effect at high pressure and at low pressure. This is $\frac{1}{2}(1 - 5) = -2$. Note that this is the same as half the difference between the pressure effect at high temperature and at low temperature.

The second major advantage of complete factorial experiments is that they are the most efficient way of estimating main effects even if no interaction is present. Suppose we are interested in the effects of two factors A and B, on a given response variable. Factor A is tested at levels A_0 and A_1, and factor B at B_0 and B_1. The simplest one-at-a-time experiment consists of one test at $(A_0 B_0)$, $(A_0 B_1)$ and $(A_1 B_0)$. Denoting the observations by the appropriate small letters we have $(a_1 b_0 - a_0 b_0)$ is an estimate of the main effect of A and $(a_0 b_1 - a_0 b_0)$ is an estimate of the main effect of B. Each of these estimates is based on just two observations and so, since there may be experimental error, it is advisable to duplicate each observation and find the average differences. We then have a total of six observations and each estimate of a main effect is based on four of these.

The simplest factorial experiment consists of one observation at each of $(A_0 B_0)$, $(A_0 B_1)$, $(A_1 B_0)$ and $(A_1 B_1)$ – a total of four observations. However both $(a_1 b_0 - a_0 b_0)$ and $(a_1 b_1 - a_0 b_1)$ are estimates of the main effect of A, and the average of these two quantities will have the same precision as the estimate from the one-at-a-time experiment, since both estimates are based on four observations. Similarly the average of $(a_0 b_1 - a_0 b_0)$ and $(a_1 b_1 - a_1 b_0)$ will have the same precision as the earlier estimate of the main effect of B. Thus the complete factorial experiment achieves the same precision as the one-at-a-time experiment but with two less observations.

The scientist who performs a one-at-a-time experiment has to assume that there is no interaction. Unfortunately the error of assuming no interaction when there really is one is perhaps the most serious industrial mistake. It is induced by the desire to perform as few tests as possible, but this is obviously a false economy if the results are misleading. In any case we have now shown that the complete factorial experiment is a more efficient method for estimating main effects, whether or not an interaction is present.

11.3 The design of complete factorial experiments

We will concentrate here on experiments involving two factors since it is a fairly straightforward matter to extend the discussion to a greater number of factors. If factor A is investigated at r levels and factor B at c levels, we have an $r \times c$ complete factorial experiment. If each treatment combination is replicated n times, the total number of tests required is given by $r \times c \times n$. As in comparative experiments, randomization should be used whenever possible, either by assigning the experimental units randomly to the treatment combinations or by performing the tests in a random order. If blocking is required, it is often convenient to let one replication of all treatment combinations form a block.

In order to illustrate some of the practical problems involved we will consider an extended version of one of the experiments considered in Chapter 10. Suppose we want to design an experiment to compare two new burners B_1 and B_2 with a standard burner B_3, and also to test the effect of a new fuel additive on the efficiency of a turbo-jet engine.

The fuel will be tested in two different proportions – low and high. These two proportions denoted by F_2 and F_3, together with the absence of the fuel additive – denoted by F_1 – are the three possible levels for the fuel factor. The instrumentation has been developed over several years and it is known that if two tests are made on each treatment combination, then the average of the two results will be sufficiently precise to detect the smallest difference considered to be of practical significance.

We have already seen that a complete factorial experiment is superior to a classical one-at-a-time test. Nevertheless it will be instructive to see the dangers of the one-at-a-time test in this particular case. The test plan for such an experiment is shown below.

Design one
One-at-a-time test

Test number	Burner	Fuel
1, 2	Standard (B_3)	Standard (F_1)
3, 4	Burner 1 (B_1)	Standard (F_1)
5, 6	Burner 2 (B_2)	Standard (F_1)
7, 8	Best burner	Low additive (F_2)
9, 10	Best burner	High additive (F_3)

Each pair of results is averaged and the best of the three burners with no fuel additive is selected for the last four tests in order to try out the fuel additive.

Two important criticisms of this design can be made. Firstly, because of the sequential nature of the experiment, it is not possible to randomize the order of the tests and so a nuisance factor may systematically affect the results. Secondly this design cannot detect interactions between the burners and the fuel, so there is a distinct possibility that the experiment will not find the best burner–fuel combination.

Two possible relationships between the burner efficiency, the burners and the fuel additive are shown in Figure 61. In the first graph the response curves are parallel so that there is no burner–fuel interaction. In this case the one-at-a-time test would find the best burner–fuel combination. But in the second graph it is clear that the effect of the fuel additive depends on the type of burner employed. In this case the one-at-a-time test would select $F_1 B_3$ as the best combination while in fact $F_2 B_2$ is better. Since it is well known that there is usually an

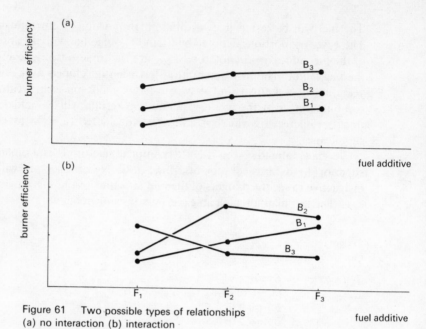

Figure 61   Two possible types of relationships
(a) no interaction (b) interaction

optimum burner for a particular type of fuel (in other words the design of the burner interacts with the type of fuel), it would be quite wrong to employ a one-at-a-time test in this case.

The second design we will consider is a randomized complete factorial experiment. With two factors each at three levels there are nine treatment combinations. Two observations are made at each treatment combination so that eighteen tests are required. The order of these tests is completely randomized and a typical test sequence is shown below.

Design two
Fully randomized factorial experiment

		Burner		
		B_1	B_2	B_3
Fuel	F_1	5, 2	18, 3	11, 14
	F_2	15, 13	1, 10	6, 16
	F_3	4, 7	12, 9	17, 8

(Numbers indicate the test sequence.)

In order to insure against severe trends in the data caused by uncontrolled factors, it is a good idea to plot the results in the order in which they are taken. This gives what is called a time-sequence plot. If an obvious trend is visible in the results its cause should be found and if possible removed, after which the experiment can be repeated. For example, eighteen tests at maximum turbine inlet temperature may be beyond the life-span of a newly designed experimental engine, so as the engine deteriorates the values of the response variable will systematically decrease.

In Design Two the reader will notice that both measurements on $B_1 F_1$ are performed early in the experiment while both experiments on $B_1 F_2$ happen to be performed much later in the experiment. If there is a trend in the results, this would mean that $B_1 F_1$ has an unfair advantage over $B_1 F_2$. This potential difficulty can be partly overcome by blocking. Divide the experiment into two blocks of nine tests and make one observation on each treatment combination in each block. A design of this type is called a *randomized block factorial experiment*. A typical test sequence is shown below.

Design three
Randomized block factorial experiment

		Burner		
		B_1	B_2	B_3
Fuel	F_1	5, 13	3, 18	7, 14
	F_2	2, 15	1, 10	6, 16
	F_3	4, 11	9, 12	8, 17

Tests 1 to 9 are in the first block and tests 10 to 18 in the second block. A similar design may also be used if the test programme is carried out on two engines or with two batches of fuel. Within each block the order of the tests is randomized and we have a randomized block factorial experiment. This is the best design for the particular experiment we have described, though a simple randomized factorial experiment will sometimes be adequate in other situations.

The data from the above experiment is given later in Example 1, and will be investigated after we have discussed the analysis of a factorial experiment in general terms.

11.4 The analysis of a complete factorial experiment

In this section we discuss the analysis of complete factorial experiments, concentrating, as in section 11.3, on two-factor experiments. The first step in such an analysis is to plot the results as shown in Figure 61, since useful information can often be obtained simply by looking at the resulting graph. For example, if the lines are approximately parallel, then there is no evidence of an interaction between the factors, whereas if the lines are skewed to one another then an interaction is probably present. Note that if one of the factors is quantitative (for example, the amount of fuel additive), this factor should be chosen as the horizontal axis in order to show up a functional relationship with the response variable.

If there is no evidence of an interaction, it is useful to calculate the main effects of the two factors by tabulating the results in a two-way table and calculating the row and column averages. These can be compared with the over-all average. However if an interaction is present, these main effects cease to have much meaning by themselves as the effect of one factor will depend on the level of the other factor.

Occasionally the above plotting and tabulation will be sufficient since the conclusions will be 'obvious'. For example, if the difference between two observations on the same treatment combination is 'small' compared with the differences between observations on different treatment combinations, then it may be possible to find the best burner–fuel combination by inspection. However the residual variation will usually be sufficiently large to make it impossible to come to such a clear-cut conclusion. In any case the experimenter is often concerned with estimating the treatment differences rather than just finding the best treatment combination. Thus a mathematical model for a factorial experiment will be proposed, after which the factor effects can be estimated. Then these effects will be tested to see if they are significantly large.

Let us suppose that factor A is investigated at r levels and factor B at c levels and that the experiment is replicated n times. The following model is proposed to describe this situation.

$$x_{ijk} = \mu + A_i + B_j + (A \times B)_{ij} + \varepsilon_{ijk}$$

$$(i = 1, \ldots, r \qquad j = 1, \ldots, c \qquad k \doteq 1, \ldots, n),$$

where μ = over-all average,

$\quad A_i$ = effect of A at ith level,

$\quad B_j$ = effect of B at jth level,

$(A \times B)_{ij}$ = joint influence of A at ith level and B at jth level; that is the interaction effect

ε_{ijk} = random error.

Since μ is the over-all average it can be shown that

$$\sum_i A_i = \sum_j B_j = 0.$$

It can also be shown that

$$\sum_i (A \times B)_{ij} = 0 \qquad \text{(for all } j)$$

and

$$\sum_j (A \times B)_{ij} = 0 \qquad \text{(for all } i).$$

It is again convenient to assume that the errors are normally distributed with mean zero and constant variance σ^2, and successive errors are independent.

The data can be tabulated in the following way.

Table 20 — Factor B

		Level 1	Level 2	$\cdots$ Level c	Row total	Row average
Factor A	Level 1	$x_{111}, \ldots, x_{11n}$	$x_{121}, \ldots, x_{12n}$	$x_{1c1}, \ldots, x_{1cn}$	$T_1.$	$\bar{x}_1.$
	Level 2	$x_{211}, \ldots, x_{21n}$	$x_{221}, \ldots, x_{22n}$	$x_{2c1}, \ldots, x_{2cn}$	$T_2.$	$\bar{x}_2.$
	$\vdots$					
	Level r	$x_{r11}, \ldots, x_{r1n}$	$x_{r21}, \ldots, x_{r2n}$	$x_{rc1}, \ldots, x_{rcn}$	$T_r.$	$\bar{x}_r.$
	Column total	$T_{.1}$	$T_{.2}$	$T_{.c}$		
	Column average	$\bar{x}_{.1}$	$\bar{x}_{.2}$	$\bar{x}_{.c}$		

The following quantities are calculated as shown:

$$\bar{x}_i. = \frac{T_i.}{nc} \qquad \bar{x}_{.j} = \frac{T_{.j}}{nr}.$$

$$T = \sum_i T_i. = \sum_j T_{.j},$$

$$\bar{x} = \frac{T}{nrc},$$

$T_{ij} = $ (sum of observations in (i, j)th cell),

$\bar{x}_{ij} = $ (average observation in (i, j)th cell)

$$= \frac{T_{ij}}{n}.$$

It can be shown that the best unbiased estimates of the model parameters are given by:

$$\hat{\mu} = \bar{x},$$

$$\hat{A}_i = \bar{x}_{i.} - \bar{x} \qquad \hat{B}_j = \bar{x}_{.j} - \bar{x},$$

$$\widehat{(A \times B)}_{ij} = \bar{x}_{ij} - \bar{x}_{i.} - \bar{x}_{.j} + \bar{x}.$$

The next step is to test the main effects of the two factors and the interaction effect to see if any of them are significantly large. The best way of doing this is by an analysis of variance. The total sum of squares, $\sum (x_{ijk} - \bar{x})^2$, is partitioned into four components. Two of these components, the row sum of squares and the column sum of squares, have the same value as the two-way ANOVA described in section 10.11. In addition the interaction sum of squares is calculated. The required formulae are given below.

Table 21
Two-factor ANOVA with Interaction

Source	Sum of squares	d.f.	E (mean square)
Main effect A (rows)	$nc \sum_{i} (\bar{x}_{i.} - \bar{x})^2$	$r-1$	$\sigma^2 + nc \sum \dfrac{A_i^2}{r-1}$
Main effect B (columns)	$nr \sum_{j} (\bar{x}_{.j} - \bar{x})^2$	$c-1$	$\sigma^2 + nr \sum \dfrac{B_j^2}{c-1}$
Interaction	$n \sum_{i,j} (\bar{x}_{ij} - \bar{x}_{i.} - \bar{x}_{.j} + \bar{x})^2$	$(r-1)(c-1)$	$\sigma^2 + n \sum \dfrac{(A \times B)_{ij}^2}{(r-1)(c-1)}$
Residual	$\sum_{i,j,k} (x_{ijk} - \bar{x}_{ij})^2$	$rc(n-1)$	σ^2
Total	$\sum_{i,j,k} (x_{ijk} - \bar{x})^2$	$rcn-1$	

The observed mean squares of the different effects are obtained by dividing the appropriate sum of squares by the appropriate number of degrees of freedom. The expected values of these mean squares are shown above. It is clear from these quantities that the A, B, and interaction mean squares should be compared with the residual mean square by means of an F-test. For example, to test the hypothesis

$$H_{01}: \text{all } A_i = 0,$$

calculate

$$F = \frac{A \text{ mean square}}{\text{Residual mean square}}.$$

If this exceeds $F_{0\cdot05, r-1, rc(n-1)}$, then H_{01} is rejected at the 5 per cent level. A similar procedure is adopted to test the hypotheses

$$H_{02}: \text{all } B_j = 0 \qquad H_{03}: \text{all } (A \times B)_{ij} = 0.$$

We will not attempt to derive the algebraic quantities shown in Table 21. However it is worth stressing that the total sum of squares is equal to the sum of the other sums of squares. Similarly the total number of degrees of freedom is equal to the sum of the constituent degrees of freedom.

The computation proceeds as follows. Calculate the row totals, the column totals, the individual cell totals and the over-all total. Also calculate $\sum\limits_{i,j,k} x_{ijk}^2$ and the correction factor T^2/nrc.

$$(\text{Total corrected sum of squares}) = \sum (x_{ijk} - \bar{x})^2$$

$$= \sum x_{ijk}^2 - \frac{T^2}{nrc}.$$

$$(\text{Row sum of squares}) = nc \sum_{i=1}^{r} (\bar{x}_{i.} - \bar{x})^2$$

$$= \frac{\sum\limits_{i=1}^{r} T_{i.}^2}{nc} - \frac{T^2}{nrc}.$$

$$(\text{Column sum of squares}) = nr \sum_{j=1}^{c} (\bar{x}_{.j} - \bar{x})^2$$

$$= \frac{\sum\limits_{j=1}^{c} T_{.j}^2}{nr} - \frac{T^2}{nrc}.$$

$$(\text{Residual sum of squares}) = \sum (x_{ijk} - \bar{x}_{ij})^2$$

$$= \sum x_{ijk}^2 - \frac{\sum\limits_{i,j} T_{ij}^2}{n}.$$

The interaction sum of squares can then be obtained by subtraction.

It is now a simple matter to calculate the mean squares and the F-ratio and see if the main effects and the interaction effect are significant.

Example 1

Two replications were made in the experiment to compare three burners and to investigate a new fuel additive. The data is given below.

Burners

		B_1	Cell total	B_2	Cell total	B_3	Cell total	Row total
Fuel	F_1	16	34	19	36	23	43	113
		18		17		20		
	F_2	19	39	25	48	19	37	124
		20		23		18		
	F_3	22	43	21	45	19	36	124
		21		24		17		
Column total			116		129		116	$T = 361$

$n = 2, r = 3, c = 3, \sum x_{ijk}^2 = 7351$ and $T^2/18 = 7240 \cdot 0$.

The ANOVA table is given below.

	Source	Sum of squares	d.f.	Mean square	F
Main effect	Fuel	13·5	2	6·8	3·3
Main effect	Burners	18·8	2	9·4	4·5
	Interaction	60·2	4	15·0	7·3
	Residual	18·5	9	2·1	
	Total	111	17		

$F_{0 \cdot 05, 2, 9} = 4 \cdot 26 \qquad F_{0 \cdot 05, 4, 9} = 3 \cdot 63$.

The interaction and burner effects are both significant at the 5 per cent level but the main effect of fuel is not significantly large. This does not mean that the fuel factor does not affect the response variable. Since the interaction is significantly large, the effect of a particular burner depends on the level of the fuel factor. We shall see how to interpret these results in the next section.

In the above example the block effect is included in the residual variation since there is no evidence of block to block variation. Let $T_{..k}$ = sum of observations in kth block. Then in Example 1, $T_{..1} = 183$ and $T_{..2} = 178$, and these values are close together. Generally with n replications, the quantity $\sum (x_{ijk} - \bar{x}_{ij})^2$ can be split into two components

$$\sum_{i,j,k} (x_{ijk} - \bar{x}_{ij})^2 = \sum_{i,j,k} (x_{ijk} - \bar{x}_{..k} + \bar{x} - \bar{x}_{ij})^2 + rc \sum_{k=1}^{n} (\bar{x}_{..k} - \bar{x})^2.$$

The first component has $(rc-1)(n-1)$ d.f. and measures the residual variation. The second component has $(n-1)$ d.f. and measures the block effect, and can be calculated with the following formula

$$rc \sum_{k=1}^{n} (\bar{x}_{..k} - \bar{x})^2 = \frac{\sum_{k=1}^{n} T_{..k}^2}{rc} - \frac{T^2}{nrc}.$$

Then the true residual sum of squares can be calculated by subtracting this quantity from $\sum_{i,j,k} (x_{ijk} - \bar{x}_{ij})^2$. Note that once we have separated the block effect, we have actually analysed a three-factor situation since the different blocks can be thought of as the levels of a third factor.

One other point of interest is that if only one replication is made (that is, $n = 1$), then, by referring to Table 21, it can be seen that the total degrees of freedom are exhausted by the main effect and interaction degrees of freedom. The formulae are then exactly the same as in a randomized block experiment (see Table 18), except that the residual sum of squares becomes the interaction sum of squares. Thus it is necessary to replicate the experiment in order to estimate the residual variation.

11.5 Follow-up procedure

The analysis of variance is a straightforward method for deciding if the main effects and/or the interactions are significantly large. But the follow-up procedure of interpreting the results is by no means so clear cut.

We begin by discussing the situation where there is no interaction. In this case the two factors can be examined separately. If one of the factors is qualitative and has a significant main effect, it is possible to find the level or levels which give the largest values of the response variable. This can be done, as in Chapter 10, with a least significant difference test. If one of the factors is continuous and has a significant main effect, the problem is to find the functional relationship between the response variable and the factor. In other words we have a regression problem.

The true type of functional relationship is usually unknown but it is often convenient to fit a polynomial to the data of the form

$$y = a_0 + a_1 x + a_2 x^2 + \ldots .$$

If the factor is investigated at c levels, it is possible to fit a polynomial up to degree $(c-1)$, although linear and quadratic terms are often all that are required. Note that the number of independent parameters involved in a polynomial model of degree $(c-1)$ is the same as the number of independent parameters involved in a model of type 10A or 10B. However if a polynomial whose degree is less than $(c-1)$ fits the data adequately then fewer parameters will be required in the polynomial model. In any case the polynomial representation will be more meaningful for a continuous variable.

One way of finding the lowest degree polynomial which adequately fits a set of data is by the method of orthogonal polynomials (see section 8.7). The total sum of squares is split up into a linear component, a quadratic component, and so on. In contrast, the main effect sum of squares given in Table 21 does not take into account the order of the levels of the factor. This means, for example, that a linear trend in the response variable may be overlooked if the non-linear effects are small, since the main effect sum of squares in Table 21 is divided by $(c-1)$ rather than one, and so may not give a significant result. The reader should be aware of such a possibility and be prepared to evaluate the different components of the main effect sum of squares (see, for example, Davies, 1956).

We now turn our attention to the situation where there is an interaction between the two factors. In this case the main effects cease to have much meaning by themselves, since the effect of one factor depends on the level of the other factor. In particular a factor cannot be discounted because its main effect is small if an interaction is present. If one of the factors is qualitative the results should be examined at each level of this factor. If the other factor is continuous, the functional relationship between the response variable and this continuous variable will depend on the level of the qualitative variable. Alternatively if both factors are continuous variables, then we have a multiple regression problem with two controlled variables.

11.6 The 2^n factorial design

A special type of complete factorial experiment is one in which n factors are each investigated at just two levels. There are then 2^n possible treatment combinations. Such an experiment is useful when a large number of factors have to be considered since it would require too many tests to run each factor at more than two levels. An experiment such as this which picks out the important factors from a number of possibilities is often called a *screening* experiment.

For simplicity we begin by considering an experiment involving just three factors A, B and C, each of which is investigated at two levels; high and low. It is useful to have a systematic method of designating the treatment combinations in such an experiment. We will use the appropriate small letter a, b and c when the corresponding factor is at the high level. The absence of a letter means that the corresponding factor is at the low level. Thus the treatment combination ab is the one in which A and B are at the high level but C is at the low level. The symbol (1) is used when all factors are at the low level. The eight possible treatment combinations are shown in Figure 62.

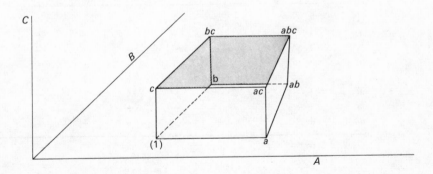

Figure 62   The eight possible treatment combinations in a 2^3 experiment

If just one observation is made at each treatment combination we have a total of eight observations. These results can be combined in seven different ways to estimate different effects. For example, there are four observations at the high level of A and four observations at the low level. The average difference between them is an estimate of the main effect of A. Denoting the observations by the corresponding treatment combination we have

$$(\text{main effect of } A) = \tfrac{1}{4}[a+ab+ac+abc-(1)-b-c-bc].$$

Similarly the main effect of B is given by

$$\tfrac{1}{4}[b+ab+bc+abc-(1)-a-c-ac].$$

An estimate of the AB interaction is given by half the difference between the main effect of A, at the high level of B, and the main effect of A, at the low level of B. Thus we have

$$(AB \text{ interaction}) = \tfrac{1}{2}[\tfrac{1}{2}(ab+abc-b-bc)-\tfrac{1}{2}(a+ac-(1)-c)]$$
$$= \tfrac{1}{4}[ab+abc+(1)+c-b-bc-a-ac].$$

Similar expressions can be found to estimate the main effect C, the AC and BC interactions and also the ABC interaction. Each of these estimates, which are called *contrasts*, is a simple linear combination of the eight observations.

It is instructive to tabulate these contrasts in the following way

Table 22

	Effect							
	Average	*A*	*B*	*AB*	*C*	*AC*	*BC*	*ABC*
(1)	+	−	−	+	−	+	+	−
a	+	+	−	−	−	−	+	+
b	+	−	+	−	−	+	−	+
ab	+	+	+	+	−	−	−	−
c	+	−	−	+	+	−	−	+
ac	+	+	−	−	+	+	−	−
bc	+	−	+	−	+	−	+	−
abc	+	+	+	+	+	+	+	+
Factor	$\frac{1}{8}$	$\frac{1}{4}$	$\frac{1}{4}$	$\frac{1}{4}$	$\frac{1}{4}$	$\frac{1}{4}$	$\frac{1}{4}$	$\frac{1}{4}$

Each effect is obtained by adding all the observations which have a plus in the appropriate column and subtracting all the observations which have a minus in this column. The result is divided by four. The calculations are completed, as shown in the first column of the table, by adding up all the observations to obtain the over-all average and dividing by eight.

Students with a knowledge of matrix theory will recognize that the columns are *orthogonal*. In simple terms this means that each of the last seven columns contain four plus and four minus signs, and if any two columns are cross-multiplied in pairs, using the rule $(+) \times (+) = (-) \times (-) = +$ and $(+) \times (-) = -$, then four plus and four minus signs will result. The practical result of this orthogonality is that the estimate of one effect will not be affected by changes in any of the other effects.

Generally it is always a good idea to see that an experimental design has the property of orthogonality. In particular a complete factorial experiment is orthogonal if each treatment combination is tested the same number of times so that the experiment is symmetric with regard to all the factors. Orthogonal designs have several advantages. Firstly the resulting calculations are simplified. Secondly the estimate of one effect is not affected by changes in one or more of the other effects. Thirdly they are efficient in the sense that for a given number of tests the effects can be estimated with greater precision using an orthogonal design than with any other type of design. The commonest form of non-orthogonal data arises when one or more observations are missing from a complete factorial experiment. Such data is much more difficult to analyse.

Another interesting point about Table 22 is that all the interaction columns can be obtained from the A, B and C columns by multiplying the appropriate columns. For example, the AB column can be obtained by multiplying the A and B columns.

An important feature of the 2^n experiment is that an analysis of variance is easy to carry out once the effect totals have been calculated. Each effect total is squared and divided by 2^n and this gives the sum of squares corresponding to that effect. A systematic method of estimating the effects and of performing the analysis of variance has been proposed by Yates. List the treatment combinations in a systematic way (see Table 22 and Table 23) and beside them list the corresponding observations. With n factors, n columns have to be calculated. Each column is generated from the preceding column in the same way. The first 2^{n-1} numbers in a column are the sums of successive pairs of numbers in the preceding column. The next 2^{n-1} numbers are the differences of successive pairs in the preceding column; the first number in a pair being subtracted from the second number. Then the final column gives the effect totals corresponding to the particular treatment combination. The effects can be estimated by dividing the effect totals by 2^{n-1}. The sum of squares for each effect is obtained by squaring the effect total and dividing by 2^n.

After calculating the sum of squares for each effect, the next problem is to see which of the effects are significantly large. In order to do this we need an estimate of the residual variation. Unfortunately we have already seen that if a factorial experiment is only replicated once, then the main effects and interactions account for all the degrees of freedom. Thus in a 2^4 experiment there are fifteen degrees of freedom which are allocated as follows.

	d.f.
Main effects	4
Two-factor interactions	6
Three-factor interactions	4
Four-factor interactions	1
Total	15

One way round this difficulty is to use the fact that interactions involving more than two factors are rarely of practical significance. Thus the residual mean square can be estimated by combining the sums of squares of interactions involving three or more factors and dividing by the corresponding number of degrees of freedom. The main effects and two-factor interactions each have one d.f., and so the mean square of each of these effects is the same as the corresponding sum of squares. A series of F-ratios can now be obtained by dividing the main effect and two-factor interaction mean squares by the residual mean square. This enables us to determine which effects are significantly large.

Example 2

The following data are the results from a 2^4 factorial experiment.

		A_1		A_2	
		B_1	B_2	B_1	B_2
C_1	D_1	18	16	12	10
	D_2	14	15	8	9
C_2	D_1	16	19	12	13
	D_2	14	11	7	8

Use Yates's method to estimate the effects. Hence perform an analysis of variance and see which effects are significantly large.

Table 23

	Observation	I	II	III	Effect total	Effect	Effect S.S.
(1)	18	30	56	116	202	12·6	
a	12	26	60	86	−44	−5·5	121·0
b	16	28	46	−22	0	0·0	0·0
ab	10	32	40	−22	2	0·25	0·25
c	16	22	−12	0	−2	−0·25	0·25
ac	12	24	−10	0	4	0·5	1·0
bc	19	21	−12	−2	4	0·5	1·0
abc	13	19	−10	4	2	0·25	0·25
d	14	−6	−4	4	−30	−3·75	56·25
ad	8	−6	4	−6	0	0·0	0·0
bd	15	−4	2	2	0	0·0	0·0
abd	9	−6	−2	2	6	0·75	2·25
cd	14	−6	0	8	−10	−1·25	6·25
acd	7	−6	−2	−4	0	0·0	0·0
bcd	11	−7	0	−2	−12	−1·5	9·0
abcd	8	−3	4	4	6	0·75	2·25

The three and four factor interaction sums of squares are combined to give

(residual sum of squares) $= 0·25 + 2·25 + 0·0 + 9·0 + 2·25$

$$= 13·75.$$

This quantity is based on five degrees of freedom so that

(residual mean square) $= \dfrac{13·75}{5} = 2·75.$

The F-ratios are as follows:

Effect	F-ratio
A	44·0
B	0·0
C	0·1
D	20·5
AB	0·1
AC	0·4
AD	0·0
BC	0·4
BD	0·0
CD	2·3

$F_{0·05,1,5} = 6·61$ $F_{0·01,1,5} = 16·26$

We conclude that the A and D effects are highly significant but that none of the other effects are significant.

11.7 Fixed effects and random effects

So far we have only considered models which are often referred to as *fixed-effects* models. For example, in the experiment to compare three burner designs, the effects of these burners can be considered to be fixed since the experimenter is interested in these particular designs and in no others. However there are some situations in which it is desirable to try and draw conclusions about a wider population than that covered by the experiment. For example, suppose we want to compare the performance of a chemical plant on different days. Then it is reasonable to consider the days on which tests were made as a random sample from the population of all possible days. In this case the factor 'days' is called a *random factor* and the corresponding mathematical model is called a *random-effects* model, or *components-of-variance* model. Using a one-way classification the random-effects model is as follows:

$$x_{ij} = \mu + t_i + \varepsilon_{ij} \qquad (i = 1, \ldots, c; j = 1, \ldots, n),$$

where $E(t_i) = 0$, and the experimental values of the t_i are a random sample from a normal distribution with mean zero and variance σ_t^2. Also assume that ε_{ij} are independent $N(0, \sigma^2)$. The fixed-effects model looks somewhat similar but the treatment effects, $\{t_i\}$, are fixed and subject to the restriction $\sum t_i = 0$.

The fixed-effects model is concerned with testing the hypothesis

$$H_0 : \text{all } t_i = 0.$$

Whereas the random-effects model is concerned with testing the hypothesis

$$H_0 : \sigma_t^2 = 0.$$

With a one-way classification it turns out that the method of analysing the data is the same. In the one-way ANOVA it can be shown that

$$E \text{ (treatment mean square)} = \begin{cases} \sigma^2 + n \sum \dfrac{t_i^2}{c-1} & \text{fixed effects,} \\ \\ \sigma^2 + n\sigma_t^2 & \text{random effects.} \end{cases}$$

In the fixed-effects model the quantity $\sum t_i^2/(c-1)$ can be thought of as measuring the spread of the t_is, and is therefore analogous to σ_t^2.

However although the analysis is the same in both cases, the distinction between the two types of model is still important because the hypotheses tested are not the same.

The distinction between the two types of model is even more important with more than one factor, since the method of analysing the data may be different. For example, if there is a significant interaction in a two-way analysis of variance, the factor mean squares must be compared, not with the residual mean square, but with the interaction mean square. The reader is referred to Wine (1964) and Rickmers and Todd (1967). Also note that it is possible to have a *mixed* model in which some of the factors are fixed and some are random.

11.8   Other topics

The material discussed up to this point covers the most important aspects of experimental design. To round off our discussion we will briefly mention several other topics, the details of which can be obtained from the references.

11.8.1   *Nested designs*

In a complete factorial experiment a test is made at every possible treatment combination, and for this reason such a design is often called a *crossed* design. However it is sometimes impossible or impractical to do this and then a *nested* design may be appropriate. For example, suppose that samples of a particular chemical are sent to four different laboratories. Within each laboratory two technicians each make two measurements on the percentage of iron in the chemical. This situation can be represented diagrammatically as follows:

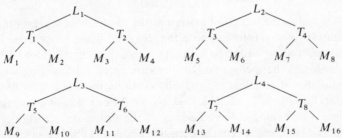

(*L* stands for laboratory, *T* for technician and *M* for measurement.)
It is obviously impractical to insist that each technician should make two measurements in each laboratory and so we say that the factor 'technicians' is nested within the factor 'laboratories'.

The following is a suitable model:

$$x_{ijk} = \mu + s_i + t_{j(i)} + \varepsilon_{ijk} \qquad (i = 1, 2, 3, 4; j = 1, 2; k = 1, 2),$$

where $x_{ijk} = k$th observation by the jth technician in the ith lab, s_i = effect of the ith lab and $t_{j(i)}$ = effect of the jth technician within the ith lab. One result of this nested design is that it is impossible to detect a technician–laboratory interaction.

11.8.2 Confounding

In a complete factorial experiment, the number of tests which form a homogeneous group may be less than the number of tests required to perform one replicate of the experiment. If higher order interactions are of little importance, a technique called *confounding* can be used to divide the tests into smaller blocks.

11.8.3 Fractional factorials

In the initial stages of an investigation, a complete factorial experiment may require too many tests. For example, with seven factors the simplest experiment of this type would require $2^7 = 128$ tests. If we are only interested in main effects and low-order interactions, it is possible to estimate these quantities by choosing a suitable fraction of the possible treatment combinations. For example a half replicate of a 2^7 experiment would require $2^{7-1} = 64$ tests. Such a design is called a *fractional factorial* design.

11.8.4 Analysis of covariance

This technique is a combination of the methods of regression and the analysis of variance. Suppose that a randomized block experiment is carried out in order to estimate a number of treatment effects. It sometimes happens that the experimenter wishes to make these estimates after adjusting the observations for the effects of one (or more) continuous variables which have been measured at the same time. For example, the efficiency of the burner of a turbo-jet engine depends on the ambient temperature, and so it would be wise to estimate the differences between burners after taking the ambient temperatures into account. The additional variable is often called a *concomitant* variable. If an analysis of variance is performed on the original data without adjustment, and if the response variable really does depend on the concomitant variable, then the results of the analysis will be inaccurate.

The simplest case is one in which the response variable, y, depends linearly on the concomitant variable, x, in which case a suitable model is as follows:

$$y_{ij} = \mu + b_i + t_j + a(x_{ij} - \bar{x}) + \varepsilon_{ij},$$

where y_{ij} = observation on treatment j in block i,

x_{ij} = corresponding value of the concomitant variable.

11.9 The examination of residuals

It should be clear by now that the standard procedure for analysing data is to propose a mathematical model to describe the physical situation, to estimate the unknown parameters of this model, and hence to draw conclusions from the data. In order that the conclusions should be valid, it is important that the model should be reasonably accurate. Thus the assumptions on which the model is based should be carefully checked. This can be done by looking at the *residuals* (see Anscombe and Tukey, 1963). The residual of an observation is the difference between the observation and the value predicted after fitting the model.

residual = observation − fitted value.

In section 10.9 we discussed how to check the assumptions involved in the mathematical model which was proposed to describe a simple comparative experiment. A similar technique can be applied to many other situations, including analysis of variance and regression models.

A preliminary examination of the residuals should be sufficient to detect any gross errors in the models. For example, if a straight line is mistakenly fitted to pairs of observations which actually follow a quadratic relationship, then a series of positive residuals will be followed by a series of negative residuals and vice versa. When the residuals are clearly non-random it may be that the type of model is incorrect or that the parameters of the model have been estimated incorrectly. For example, if an arithmetical mistake has been made we might find that all the residuals have the same sign.

Another useful procedure is to plot the residuals against external uncontrolled variables. For example, if the tests are made sequentially, the residuals can be plotted against time. If they do not appear to be random, the experimenter should try to find the cause of the non-randomness and if possible remove it.

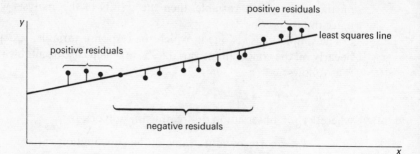

Figure 63

Another important use of residuals is to detect *outliers* in a set of data. These are 'wild' observations which do not seem to be consistent with the rest of the data. One common procedure is to reject observations whose residual is more than four times the residual standard deviation. Another technique is to plot the residuals on normal probability paper and look at the largest and smallest observations. If an observation is suspect, it should, if possible, be checked by looking at the original records to see if there is something wrong with one of the measurements. The reader is warned that considerable care should be taken before an observation is rejected, because a large residual can be the result of fitting the wrong model.

If an outlier is rejected it can be completely removed or replaced with an average value. Alternatively, in some situations, it is reasonable to replace the outlier with the next largest (or smallest) observation. This latter technique called Winsorization, (see Dixon, 1960), is often used to calculate the sample mean when one observation is significantly higher or lower than the rest.

11.10 Determination of optimum conditions

Up to this point we have considered situations in which the objective has been to obtain a general picture of how the system is affected by changes in the controlled factors. However it sometimes happens, particularly in the chemical industry, that the prime objective is simply to find the conditions which maximize (or minimize) some performance criterion. For example, we might want to minimize the total cost per unit yield. Since a minimization problem can be converted to a maximization problem, we will only consider the latter.

Firstly it is important to understand what is meant by a response surface. The response variable, y, is an unknown function of several variables, $x_1, \ldots, x_k$:

$$y = f(x_1, \ldots, x_k).$$

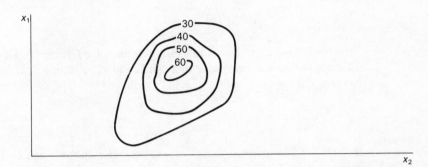

Figure 64   A contour diagram

We will consider the case where all the variables are continuous. In the case of one or two controlled variables, it is convenient to think of this function in geometrical terms. With just one variable, x_1, the relation between y and x_1 can be represented by a curve whereas with two variables, x_1 and x_2, the relation between y and x_1 and x_2 can be represented by a surface whose height, at particular values of the controlled variables, is equal to the corresponding value of the response variable. These values of the response variable generate what is called a response surface. It is often convenient to represent this response surface by a contour diagram, similar to a geographical map. On each contour the value of the response variable is a constant.

The problem of maximizing y is complicated by several factors. Firstly there may be restrictions on the values which the controlled variables can take; secondly we may not know the type of functional relationship between the response variable and the controlled variables, and thirdly there will be measurement errors in the observed values of the response variable. However in chemical processes the coefficient of variation of observations made under identical conditions is usually reasonably small.

Methods of maximizing y are discussed in Davies (1956). A straightforward though laborious procedure is as follows. With a single controlled variable, x, measure the response variable over a wide range of values of x, and plot the results on a scatter diagram. If a smooth curve is drawn through the observations, an estimate of the maximum will usually be visible. Alternatively a suitable function can be fitted to the data by the method of least squares. For example, we could assume that there is a quadratic relationship between y and x of the form

$$y = a_0 + a_1 x + a_2 x^2.$$

Estimates of a_0, a_1 and a_2 can be found as shown in Chapter 8. Then an estimate of the optimum value of x can be found by setting dy/dx equal to zero. This gives

$$\hat{x}_{opt} = -\frac{\hat{a}_1}{2\hat{a}_2}.$$

If there are k controlled variables, the response variable can be measured at a grid of points throughout the region of interest. In other words a complete factorial experiment is performed. A response surface can be fitted to the data and this enables us to estimate the optimum conditions. Unfortunately this procedure may require a large number of tests.

An alternative procedure is to adopt a sequential approach as described in Box and Wilson (1951). If there is one controlled variable, the response variable is measured at what is considered to be a reasonable starting point, say x_0, and also at a somewhat higher value $x_0 + a$. The position of the third measurement depends on the first two results. It is made above $x_0 + a$, if the second observation is the larger, and below x_0, if the first observation is the larger. However if the first two observations are about the same, the third measurement can be made between x_0 and $x_0 + a$. This step procedure continues until a maximum is found.

In the case of two controlled variables a procedure called the *method of steepest ascent* can be employed. If the initial conditions, at the point P are not too close to the maximum, the response surface can be approximated locally by a plane.

$$y = a_0 + a_1 x_1 + a_2 x_2.$$

If the experimental error is small, these parameters can be estimated with a 2^2 experiment around P. Then an estimate of the path of steepest ascent can be obtained by making changes in the $\{x_i\}$ which are proportional to the $\{\hat{a}_i\}$. The reader is warned that this path will depend on the scales in which the controlled variables are measured.

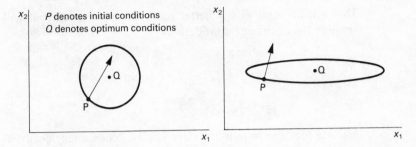

Figure 65 (a) Path of steepest ascent goes close to Q (b) Path of steepest ascent does not go close to Q

Best results are obtained if the contour lines are approximately circular; if the contour lines are elongated ellipses then the results will be rather poor. The reason for this is that the path of steepest ascent is not preserved when a linear transformation is made on one of the controlled variables. The best procedure is to standardize the controlled variable in such a way that unit changes in each variable can be expected to have about the same effect. This is best illustrated by an example.

Example 3

The percentage of the theoretical yield of a chemical process, y, depends on the temperature x_1 and the percentage of one of the chemicals, x_2. The objective of the experimental programme is to maximize y. A reasonable starting point is known to be $x_1 = 425°C$ and $x_2 = 11$ per cent. A change of 25°C in x_1 and a change of 1 per cent in x_2 are thought to be roughly equivalent. A 2^2 experiment was carried out and the results are given below.

		x_1	
		400	450
x_2	10	50	52
	12	54	56

It is possible to fit a plane to the data as it stands, but it is much better to standardize the controlled variables with the following linear transformations

$$x_1' = \frac{x_1 - 425}{25} \qquad x_2' = \frac{x_2 - 11}{1}.$$

283 Determination of optimum conditions

This will not only give a better path of steeper ascent but will also simplify the arithmetic considerably.

		x'_1	
		-1	$+1$
x'_2	-1	50	52
	$+1$	54	56

Assume that the response surface can be represented locally by the plane

$$y = a_0 + a_1 x'_1 + a_2 x'_2.$$

Then estimates of a_0, a_1 and a_2 can be obtained by the method of least squares, as shown in section 8.6. Since the experiment is orthogonal and the controlled variables have been standardized we have

$$\sum x'_{1i} = \sum x'_{2i} = \sum x'_{1i} x'_{2i} = 0.$$

This considerably simplifies the least squares normal equations to give

$$\hat{a}_0 = \bar{y} = 53,$$

$$\hat{a}_1 = \frac{\sum y_i x'_{1i}}{\sum x'^2_{1i}} = \frac{52 + 56 - 54 - 50}{4} = 1,$$

$$\hat{a}_2 = \frac{\sum y_i x'_{2i}}{\sum x'^2_{2i}} = \frac{54 + 56 - 52 - 50}{4} = 2.$$

Thus the estimated plane is given by

$$y = 53 + x'_1 + 2x'_2.$$

The path of steepest ascent can be obtained by changing x'_1 and x'_2 in the ratio $1:2$. But a change of one unit in x'_1 is equivalent to a $25°C$ change in x_1, and a change of two units in x'_2 is equivalent to a change of 2 per cent in x_2. Thus the path of steepest ascent can be obtained by starting at $x_1 = 425°C$ and $x_2 = 11$ per cent, and changing x_1 and x_2 in the ratio $25°C:2$ per cent. Some possible points for further analysis are given below.

	x_1	x_2
Starting point	425	11
Possible points	$437\frac{1}{2}$	12
	450	13
	$462\frac{1}{2}$	14
	475	15

Observations are made along this line until a maximum is found, then the procedure can be repeated.

Note that if the variables are not standardized, the least squares equations will be much more difficult to solve. In addition the response contours are very elongated ellipses in the original units whereas they will be closer to circles in the standardized units. This means that the method of steepest ascent will get to the maximum more quickly by using the standardized units.

In the second stage of the optimization procedure, as the values of the controlled variables get closer to the optimum conditions, the second order effects will become 'large' compared with first order effects. Then it will no longer be sufficient to approximate the response surface with a plane. Instead it will be necessary to perform a 3^n experiment or to use a design called a *composite* design (see Davies, 1956).

We will conclude this section by briefly mentioning another optimization procedure called *evolutionary operation*, which has been proposed in Box (1957). This should be adopted when the process is already operating at an acceptable level, and when it would be impractical to interrupt production to carry out one of the experiments already described. Evolutionary operation is often abbreviated to Evop.

Suppose there are k controlled variables—usually two or three. Then a 2^k experiment can be carried out by making small changes in these variables. These changes should be chosen so that they do not seriously affect the manufacturing process. For example, if a change of 20°C in the process temperature is known to have a significant effect then a much smaller change of say 2°C should be made. Since the changes are so small, it will be necessary to replicate the experiment several times in order to detect significant changes in the response variable. At least three replicates or cycles of the 2^k experiment will be required. Then the local properties of the response surface (main effects and first-order interactions) can be estimated. Once a significant effect has been detected, the operating conditions can be changed in the required direction, after which a new phase will be started. This procedure can be continued indefinitely.

11.11 Summary

The choice of the appropriate experimental design will depend upon a number of factors, but the most important is the state of knowledge of the phenomena or mechanism which is being tested. Many industrial experiments are carried out in situations in which there is already a considerable amount of information available. In other words the physical laws governing the behaviour of the system are well understood. In this sort of situation the statistician is usually only required to improve the precision and accuracy of the data. Thus when the state of knowledge is high, the experimental uncertainty is minimized and the need for statistics is not so acute.

Experiments which do not fall in this category are fewer in number but much more important. The true research experiment is by definition not founded on well established knowledge. For example, there are many instances in the history of science in which an experiment has given a surprising result or has led to an accidental discovery. This type of experiment, which has a high degree of experimental uncertainty, will require all the skill of both the engineer and the statistician. The experimental designs, which have been described in the last two chapters, will then be of tremendous value and considerable care should be taken to choose the most suitable design.

Exercises

1. A complete factorial experiment is set up to investigate factor A at r levels and factor B at c levels. The experiment is replicated n times. Show that

$$\sum_{i=1}^{r} \sum_{j=1}^{c} \sum_{k=1}^{n} (x_{ijk} - \bar{x})^2 = nc \sum_{i=1}^{r} (\bar{x}_{i.} - \bar{x})^2 + nr \sum_{j=1}^{c} (\bar{x}_{.j} - \bar{x})^2$$

$$+ n \sum_{i=1}^{r} \sum_{j=1}^{c} (\bar{x}_{ij} - \bar{x}_{i.} - \bar{x}_{.j} + \bar{x})^2 + \sum_{i=1}^{r} \sum_{j=1}^{c} \sum_{k=1}^{n} (x_{ijk} - \bar{x}_{ij})^2.$$

2. An experiment was set up to investigate the effect of four different catalysts, A, B, C and D, and the effect of agitating the mixture on the yield from a chemical reaction. No agitation will be denoted by E_1, medium agitation by E_2 and high agitation by E_3. A 4×3 factorial experiment, twice replicated, was carried out. One replication formed a block and the order of experiments within a block was randomized. The results were as follows (percentage of theoretical yield).

	E_1		E_2		E_3	
A	44		49		50	
		46		48		50
B	46		51		51	
		45		47		52

C	50		53		55
		52		55	54
D	46		47		50
		45		50	49

Set up the ANOVA table. Graph the results, using 'amount of agitation' as the horizontal variate. Discuss the results of the analysis.

3. A singly replicated 2^3 experiment is carried out with three factors, A, B and C. Use Yates's method to calculate the main effects and interactions from the following data.

	A_1		A_2	
	B_1	B_2	B_1	B_2
C_1	17	18	16	14
C_2	10	13	8	6

Calculate the sum of squares for each effect. Test the main effects by comparing them with the mean square obtained by combining the two- and three-factor interactions.

References

ANSCOMBE, F. J., and TUKEY, J. W. (1963), 'The examination and analysis of residuals', *Technometrics*, vol. 5, pp. 141–60.

BOX. G. E. P. (1957), 'Evolutionary operation'. *Applied Statistics*, vol. 4, pp. 3–23.

BOX, G. E. P., and WILSON, K. B. (1951), 'On the experimental attainment of optimum conditions', *Journal of the Royal Statistical Society*, Series B, vol. 13, pp. 1–45.

DAVIES, O. L. (ed.) (1956), *The Design and Analysis of Industrial Experiments*, Oliver & Boyd, 2nd edn.

DIXON, W. J. (1960), 'Simplified estimation from censored normal samples', *Annals of Mathematical Statistics*, vol. 31, pp. 385–91.

RICKMERS, A. D., and TODD, H. N. (1967), *Statistics*, McGraw-Hill.

WINE, R. L. (1964), *Statistics for Scientists and Engineers*, Prentice-Hall.

Chapter 12
Quality control

This chapter is concerned with some of the problems involved in controlling the quality of a manufactured product. The first part of the chapter (sections 12.1–7) deals with *acceptance sampling*, which is concerned with monitoring the quality of manufactured items supplied by the manufacturer to the consumer in batches. The problem is to decide whether the batch should be accepted or rejected on the basis of a sample randomly drawn from the batch.

The second part of the chapter (sections 12.8–11) is concerned with *process control*. The problem is to detect changes in the performance of the manufacturing process and to take appropriate action when necessary to control the process.

Most control schemes are usually planned by a statistician but in order to operate them it is useful for an engineer to understand the ideas behind them and to know the meaning of some of the commonly used expressions. Thus we will summarize the basic principles and provide an introduction to the more commonly used techniques. Most of the material in this chapter, with the exception of section 12.11, comes under the general heading of statistical quality control. The associated problem of life-testing is considered in the next chapter.

12.1 Acceptance sampling

Any manufacturing process will inevitably produce some defective items. The manufactured items will often be supplied by the manufacturer to the consumer in *batches* or *lots*, which may be examined by the manufacturer before shipment or by the consumer before acceptance. The inspection often consists of drawing a sample from each batch and then deciding whether to accept or reject the batch on the evidence provided by the sample. A variety of sampling schemes exist and we will describe the more important of these.

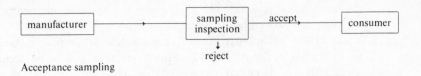

Acceptance sampling

A simple type of sampling scheme is one in which a single sample is taken and the batch is accepted if there are not more than a certain number of defective items. For example, we could take a sample size 100 from each batch and reject the batch if there is more than one defective item. Otherwise the batch is accepted.

Acceptance sampling is used when the cost of inspecting an item is such that it is uneconomic to look at every item in a batch. For example, it must be used in the case where the manufactured item is destroyed by the inspection technique. In contrast, in precision engineering it is more common to inspect every item, in which case there are few statistical problems and most of the following remarks do not apply.

When a batch is rejected by a sampling scheme, it may be returned to the manufacturer, it may be purchased at a lower price or it may even be destroyed. Alternatively rejected batches may be subjected to 100 per cent inspection so that all defective items in the batch are replaced by good items. A sampling plan of this type, called a *rectifying scheme*, is considered in section 12.4.

Acceptance sampling plans can be divided into two classes. If the items in a sample are classed simply as 'good' or 'defective', then the sampling scheme is said to be *sampling by attributes*. This qualitative approach contrasts with *sampling by variables*, in which a quantitative measurement is involved. In other words an attribute scheme does not say how good or how defective an item is. Sometimes this is inevitable. For example, a light bulb will either work or it will not work. There is no in-between state and we must use sampling by attributes.

We shall concentrate our attention on attribute sampling schemes. Fortunately many of the general principles involved also apply to sampling by variables.

12.2 Operating characteristic curve

The performance of a particular sampling plan may be described by the operating characteristic curve or O–C curve. Let p denote the proportion of defectives in a batch. If p is very small, we would like the batch to be accepted; on the other hand if p is large, we would like the batch to be rejected. The O–C curve is a graph of the probability of accepting a batch plotted against p.

Let $L(p)$ = probability of accepting a batch. When $p = 0$, there are no defectives in the batch and so the batch is certain to be accepted. Thus $L(0) = 1$. However when $p = 1$, all the items are defective and so the batch is certain to be rejected. Thus $L(1) = 0$.

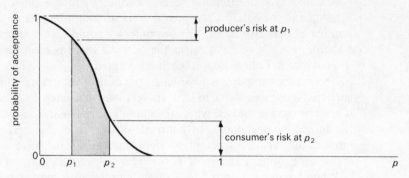

Figure 66   O–C curve

The problem is to design a sampling plan so that batches of 'good' quality are likely to be accepted whereas batches of 'bad' quality are likely to be rejected. The producer and consumer should get together and decide on a sampling plan which is fair to both. Firstly the consumer should specify the quality level which he would like the producer to achieve. The proportion of defectives in a batch which is acceptable to the consumer is called the *acceptable quality level* (abbreviated A.Q.L.) and will be denoted by p_1. Ideally we would like to be certain of accepting such a batch but this is not possible without 100 per cent inspection. The probability that a 'good' batch will be rejected because of a pessimistic looking sample is called the *producer's risk* and is denoted by α. The sampling scheme is often chosen so that α has some agreed value such as 5 per cent.

The consumer must also decide on a quality level which is definitely unacceptable. The proportion of defective items in a batch which is considered 'bad' is called the unacceptable quality level and will be denoted by p_2. The corresponding percentage, namely $100p_2$, is often called the *lot tolerance percentage defective* (abbreviated L.T.P.D.). Ideally we would like to be certain of rejecting a batch of this quality, but this is also not possible without 100 per cent inspection. The probability that a bad batch will be accepted because of an optimistic looking sample is called the *consumer's risk* and is denoted by β. The sampling scheme is often chosen so that β has some agreed value such as 10 per cent.

The producer's and consumer's risk are illustrated in Figure 66. We have said that values of p above p_2 are definitely unacceptable. Nevertheless the manufacturer should aim to do better than this and his quality control programme should attempt to keep p below p_1. If the fraction defective increases above p_1 towards p_2, the batch becomes increasingly likely to be rejected by the sampling scheme and this is financially harmful to the producer.

The observant reader will note that the operating characteristic curve is none other than the power curve when testing the hypothesis

$$H_0 : p = p_1$$

against the alternative hypothesis

$$H_1 : p > p_1 .$$

If the batch is rejected by the sampling scheme this is equivalent to rejecting the above null hypothesis. If the batch is rejected when H_0 is actually true, we have an error of type I and the probability of such an error has been specified as α. Conversely if the batch is accepted when it should be rejected, we have an error of type II. When $p = p_2$ the probability of a type II error has been specified as β.

Example 1

What is the O–C curve for a sampling scheme such as that quoted in section 12.1 in which a single sample is taken and the batch accepted if there is not more than one defective item?

Let p denote the proportion of defective items in a batch. Let us also assume that the sample size is small compared with the batch size. Then the number of defective items in a sample will be a random variable which follows the binomial distribution with parameters $n = 100$ and p. (Strictly speaking it will follow a hypergeometric distribution, see exercise 12, Chapter 4.)

Thus $P(0 \text{ defectives}) = (1-p)^{100}$,

$P(1 \text{ defective}) = 100p(1-p)^{99}$.

Thus the probability of accepting a batch is given by

$(1-p)^{100} + 100p(1-p)^{99}$.

This was calculated for various values of p and the O–C curve is shown in Figure 67.

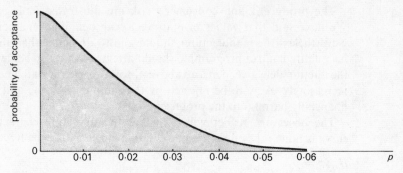

Figure 67   O–C curve for the sampling scheme given in Example 1

Note that it would be easier to use the Poisson approximation to the binomial distribution with

$$\mu = np = 100p.$$

Then the probability of accepting a batch is given approximately by

$$e^{-100p} + 100p \, e^{-100p}.$$

The O–C curve can be used to explain what is meant by an ideal sampling scheme. This scheme would be such that any batches with less than a proportion p_3 of defectives would be accepted, whereas any batches with more than a proportion p_3 of defectives would be rejected; the consumer would presumably specify p_3 somewhere between p_1

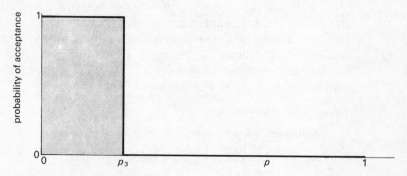

Figure 68   An ideal O–C curve

and p_2. The O–C curve would then be z-shaped as in Figure 68. But this O–C curve can only be realized with 100 per cent inspection. However the larger the sample size the closer the O–C curve will approach the ideal z-shaped curve. Generally speaking if there is only a small difference between the specified values of p_1 and p_2 then large samples will have to be taken to get an O–C curve which is sufficiently z-shaped.

It is usually a fairly straightforward task to calculate the O–C curve for a particular sampling scheme, such as that in Example 1. However in practice we may want to design a sampling scheme for given values of p_1, α, p_2 and β. For example, the consumer may want the manufacturer to aim at producing material which is 2 per cent defective or better. If a producer's risk of 5 per cent is specified, we have $p_1 = 0.02$ and $\alpha = 0.05$. In addition the consumer may decide that batches containing more than 7 per cent defective items are definitely unacceptable. If a consumer's risk of 10 per cent is specified, we have $p_2 = 0.07$ and $\beta = 0.10$. Thus two points on the O–C curve have been selected and the problem is to find a sampling scheme whose O–C curve goes through these two points.

We will now describe several types of sampling scheme and show how these are constructed given the specified values of p_1, p_2, α and β.

12.3 Types of sampling schemes

12.3.1 Single sampling

This is the simplest type of sampling plan. It consists of taking a sample size n from each batch and accepting the batch as satisfactory provided the number of defective items in the sample does not exceed a given number c. The quantity c is called the acceptance number. Example 1 deals with a single sampling scheme.

It may be impossible to find a plan which meets our requirements exactly, because the sample size and acceptance number must be integers. However a reasonable approximation may be found by consulting the tables given in Bowker and Lieberman (1957) and Duncan (1965).

12.3.2 Double sampling

A simple extension of the single sampling scheme is obtained by a two-stage sampling procedure. A sample size n_1 is drawn but the batch need not be accepted or rejected as a result of this first sample if it leaves some doubt as to the quality of the batch. Instead a second sample size n_2 can be drawn and the results of the two samples combined before a decision is made. The scheme will depend upon three constants c_1, c_2 and c_3.

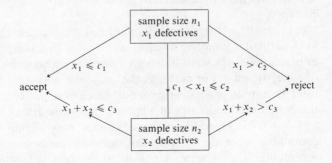

The plan is usually simplified by taking $c_2 = c_3$. As in the case of the single sampling plan, we would like to find values for n_1, n_2, c_1 and c_2 given the specified values for p_1, p_2, α and β. In fact we need a further restriction on the parameters to derive a unique double sampling plan. This is usually achieved by fixing the ratio $n_2 : n_1$; for example, we can take $n_2 = 2n_1$. The reader is referred to the tables in Columbia University (1948) and Duncan (1965).

A feature of double sampling schemes is that a very good lot will be accepted and a very bad lot rejected as a result of the first sample. This first sample is smaller than the number of items inspected in the equivalent single sampling, which has the same O–C curve. Thus double sampling enables sampling costs to be reduced as well as providing the psychological advantage of giving a batch a second chance.

12.3.3 Sequential sampling

The principle of double sampling can be extended to sequential sampling. Using this technique the sample is built up item by item, and after each observation a decision is taken on whether the batch should be accepted, rejected or whether another observation should be taken.

As before we must specify the four quantities p_1, α, p_2 and β. Then three constants h_1, h_2 and s, which characterize the chart on which the results are plotted, can be calculated. These constants are given by

$$h_1 = \frac{\log\left(\dfrac{1-\alpha}{\beta}\right)}{\log\left[\dfrac{p_2(1-p_1)}{p_1(1-p_2)}\right]},$$

$$h_2 = \frac{\log\left(\dfrac{1-\beta}{\alpha}\right)}{\log\left[\dfrac{p_2(1-p_1)}{p_1(1-p_2)}\right]},$$

$$s = \frac{\log\left(\dfrac{1-p_1}{1-p_2}\right)}{\log\left[\dfrac{p_2(1-p_1)}{p_1(1-p_2)}\right]}.$$

After each item is inspected, the cumulative sample size n and the cumulative number of defective items d are known. If $d > sn + h_2$, then the batch is rejected and if $d < sn - h_1$, the batch is accepted. Otherwise another item is inspected. Continue sampling until the batch is accepted or rejected.

Sequential sampling was developed during the last war by A. Wald (1947) and G. A. Barnard (1946); for further information see also Columbia University (1945). Note that a sequential sampling scheme is more efficient than the equivalent double sampling scheme which has the same specifications.

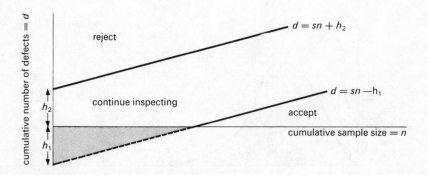

Figure 69   Sequential sampling chart

12.4 Rectifying schemes

The first major contribution to the theory of sampling inspection was made by Dodge and Romig (1944) who considered a somewhat different situation. Any batch which is rejected by the sampling scheme is subjected to 100 per cent inspection and rectification. Thus all defective items are replaced with good items.

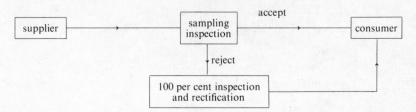

The consumer will receive two types of batches. The first type will contain some defective items, but have been accepted by the sampling inspection. The second type will contain no defective items as they have been subjected to 100 per cent inspection and rectification. From these two types of batches we can calculate the *average outgoing quality* (abbreviated A.O.Q.), which is the average proportion of defectives in batches received by the consumer.

As before, let p denote the proportion of defective items in a batch before the sampling inspection. When $p = 0$ it is clear that A.O.Q. = 0. When $p = 1$, all the batches will be subjected to 100 per cent inspection and rectification so that again we have A.O.Q. = 0. In between these two values, the A.O.Q. will have a maximum which is called the *average outgoing quality limit* (abbreviated A.O.Q.L.).

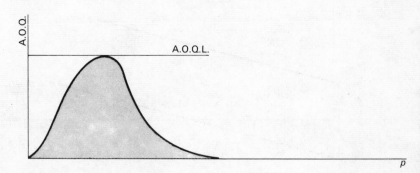

Figure 70

The advantage of a rectifying scheme is that the consumer knows that the A.O.Q. must be less than the A.O.Q.L. whatever the proportion of defectives supplied by the producer.

Dodge and Romig (1944) give tables for deriving sampling schemes in two different ways. In both types of scheme it is necessary to specify the average quality of material produced by the process. The average proportion of defectives is called the *process average percentage defective*. If this is close to the unacceptable quality level, then a large amount of sampling will be required; on the other hand if it is relatively low, then less sampling will be required. It can be estimated from past data by dividing the total number of defectives by the total number of items inspected. In the first type of sampling scheme it is necessary to specify the process average and the L.T.P.D. with a consumer's risk of 10 per cent, and in the second type of sampling scheme it is necessary to specify the process average and the A.O.Q.L. The tables mentioned above give the sampling schemes which require the minimum over-all amount of inspection to satisfy these specifications.

12.5 The military standard plan

This set of tables (U.S. Department of Defense, 1963) were originally developed during the last war but are now available for civilian use and can be obtained from the U.S. Government Printing Office. They combine many of the features of earlier plans with some new ones. Firstly defects are grouped into three different classes; critical, major and minor. This is because some defects are much more serious than others and should be weighted accordingly. For example, a paint scratch on a car is not comparable to faulty steering.

Secondly three levels of inspection are available. They are

 (i) *normal*,
 (ii) *reduced*, smaller samples than usual are taken,
 (iii) *tightened*, larger samples than usual are taken.
The choice of the level of inspection depends on how close the estimated process average is to the A.Q.L. Thus the scheme adopts the sensible approach of taking into account the quality of recent batches. If the production line has hit a bad patch then it is sensible to take larger samples than usual. On the other hand if the process has been producing good batches for a long period then reduced sampling can be employed. The sampling scheme is chosen in such a way that the producer's risk is much smaller for large lots than for small lots. The reason for this

is that it is much more serious to reject a large batch when it is 'good' than it is to reject a small batch. A description of this plan is given in Bowker and Lieberman (1957) together with tables.

12.6 Sampling by variables

If a quality characteristic is a continuous variable, it may be possible to use a variables sampling scheme rather than an attribute scheme. A quantitative measurement provides more information than a simple statement that an item is or is not defective, and so a variables sampling scheme can give the same protection, with a smaller sample size, as an attribute sampling scheme. For example, suppose that the upper specification limit for some quality characteristic is given by U. Then an attribute scheme would simply note the number of items in a sample whose value exceeded U. A variables scheme would involve finding the exact measurement for each item in the sample and calculating the sample mean $\bar{x}$. Thus if the standard deviation, σ, of the measurements is known, and it is reasonable to assume that the measurements are normally distributed, then it is easy to estimate the proportion of measurements which will exceed U. The batch will be rejected if $(U - \bar{x})\sqrt{n}/\sigma$ exceeds a certain constant. If σ is unknown it must be estimated either from the sample standard deviation s or from the sample range R. Details of such procedures can be found in Bowker and Lieberman (1957) or Duncan (1965).

One disadvantage of a variables plan is that if several features of a product are of interest then it will be necessary to make several measurements, whereas in an attribute scheme it would still be relatively simple to decide whether the item was defective.

12.7 Practical problems

We have so far said little about the practical problems involved in setting up an acceptance sampling scheme. For example, we have seen that a sequential sampling scheme is more efficient than a double sampling scheme which in turn is more efficient than a single sampling scheme. However if we consider ease of operation the order is reversed. For this reason a single sampling scheme may give more reliable results if operated by untrained personnel. More information about setting up acceptance sampling schemes can be obtained from the references.

12.8 Process control

The second major branch of statistical quality control is that of process control. This is concerned with the important problem of keeping a manufacturing process at a specified stable level, and in consequence has received widespread application in industry. It has three main ingredients:

(1) detecting changes in process performance

(2) finding the cause of these changes

(3) making appropriate adjustments to the process.

A process which has been running at an acceptable level may suddenly move off target. The job of the sampling inspector is to detect this change as quickly as possible so that appropriate corrective action can be taken.

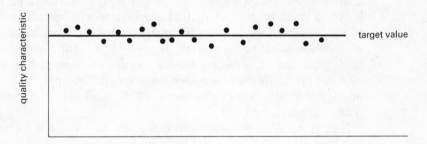

Figure 71   A control chart

The most commonly used tool for detecting changes is the *control chart* or *Shewhart chart*. Choose a variable which is characteristic of the quality of the process and plot this against time. For example, a chemical process may have a certain optimum temperature. In this case the actual temperature can be measured at regular intervals and compared with the target value by plotting it on a control chart. Of course successive measurements will vary to a greater or lesser degree even when the process is on target. Some of the causes of variation are outside the control of the manufacturer and these combine to form what is called the residual variation. However some causes of variation can be identified and removed and these are called *assignable* causes. A process is said to be under control if all the assignable causes of variation have been removed.

The use of control charts to detect changes in process performance has a history of about forty years and was pioneered by Shewhart (1931) in America and Dudding in England. The biggest impact of the control chart is visual. In other words the experienced inspector can get a lot of information simply by looking at the chart. In addition a variety of statistical techniques have been proposed to help him decide when a change has occurred.

12.8.1 Action lines

One simple device is to draw action lines on the control chart at $T \pm 3\sigma$ where T is the target value and σ is the residual standard deviation, that is, the standard deviation of successive observations when the process is under control. The action lines are sometimes called control limits. They are chosen so that if the process is on target there is only a small probability that an observation will be outside them. If the observations are approximately normally distributed, this probability should be about 0·003. However these action lines are also used when the observations are not normally distributed in which case this probability will be somewhat different. The upper action line or control limit is often abbreviated U.C.L. Similarly the lower limit is often abbreviated L.C.L.

If an observation does fall outside the control limits, it is an indication that the process has moved off target and some form of action is required. This action may consist of turning a knob, resetting a machine, replacing a faulty piece of equipment or simply taking several more measurements to check the first measurement.

Action lines are a useful guide if used intelligently. However because the observations are treated independently, the method is insensitive to small changes in the mean. For example, if the process mean moves a distance σ off target, an average of forty-four results are plotted before one falls outside the control limits.

12.8.2 Warning lines

The above method can be modified by inserting warning lines on the control chart at $T \pm 2\sigma$. Action is taken if two consecutive results lie outside the warning lines.

12.8.3 *Rule of seven*

This is a useful rule of thumb which is often used by quality control engineers. A run of seven observations on the same side of the target value is taken to indicate a change in the process mean.

12.9 Control charts for samples

So far we have considered the situation where just one observation is taken at regular intervals; however control charts can also be used when small samples are taken at regular intervals. If a manufacturing process makes a large number of items, one way of controlling the quality of the goods as they are made is to take relatively small samples in spot checks. The problems involved are distinct from those involved in acceptance sampling where much larger samples are involved. Of course a small sample is not so reliable as a large sample, but nevertheless it can provide valuable information, especially if carried out by an experienced inspector who goes round the shop floor in a fairly systematic way and who can make adjustments on the spot where necessary.

In a chemical process it may not be possible to take several observations at the same time and then it is often convenient to divide the data into what are called rational subgroups. Subgroups of four or five are commonly used, and within each subgroup the variation is assumed to be random. These subgroups can now be treated as samples.

Variables. Suppose that a sample of n measurements is taken at regular intervals on a continuous variable. Let T denote the target value and σ denote the residual standard deviation. We want to control not only the average quality of the process but also the variability within successive samples. The first of these objectives can be achieved by plotting successive sample means on a control chart called an $\bar{x}$-chart. The standard error of the sample mean is given by $\sigma/\sqrt{n}$ so that if σ is known we can draw action lines at $T \pm 3\sigma/\sqrt{n}$. The second objective is achieved by plotting the sample range, R, on a control chart called an R-chart. The sample range is much easier to calculate than the sample standard deviation and in section 2.5 we noted that the range is useful for comparing the variability in samples of equal size. Thus the range is usually preferred to the standard deviation in this situation.

If the observations are normally distributed, it can be shown that the sampling distribution of the range has mean $k_1\sigma$, and upper percentage points $R_{0.025}$ and $R_{0.001}$ given by $k_2\sigma$ and $k_3\sigma$ respectively, where k_1, k_2 and k_3 depend on the sample size. The R-chart is constructed by placing a warning line at $k_2\sigma$ and an action line at $k_3\sigma$. The target value is at $k_1\sigma$. Values of k_1, k_2 and k_3 are given in Table 14 of Pearson (1960) and are reprinted by permission in Table 24.

Table 24

n	2	3	4	5	6	7	8	9	10	11	12
k_1	1·13	1·69	2·06	2·33	2·53	2·70	2·85	2·97	3·08	3·17	3·26
k_2	3·17	3·68	3·98	4·20	4·36	4·49	4·61	4·70	4·79	4·86	4·92
k_3	4·65	5·06	5·31	5·48	5·62	5·73	5·82	5·90	5·97	6·04	6·09

So far we have assumed that σ is known; however it is usually necessary to estimate it from past data. Rather than calculate the average sample standard deviation, it will usually be simpler to calculate the average sample range, $\bar{R}$. An estimate of σ is given by $\bar{R}/k_1$ and the R-chart consists of a target value at $\bar{R}$, an upper warning line at $k_2\bar{R}/k_1$, and an upper action line at $k_3\bar{R}/k_1$.

Attributes. Suppose that a sample of n items is inspected at regular intervals. Then the number of defective items in the sample can be plotted on a control chart. The action and warning lines are calculated as follows. Suppose that on past evidence the average proportion of

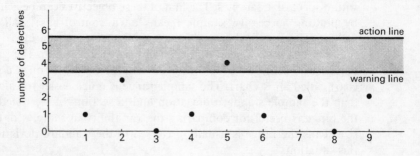

Figure 72   A typical control chart for sampling by attributes

defective items is p and that this is a satisfactory quality level. If this quality level does not change, the number of defective items in successive samples will be a binomial variable with parameters n and p. The probability of getting r defective items is given by ${}^nC_r p^r(1-p)^{n-r}$ or, using the Poisson approximation, by

$$\frac{e^{-np}(np)^r}{r!}.$$

The action limit, A, is chosen so that

$$\sum_{r \geqslant A} {}^nC_r p^r(1-p)^{n-r} \leqslant \tfrac{1}{1000}.$$

Similarly the warning limit, W, is chosen so that

$$\sum_{r \geqslant W} {}^nC_r p^r(1-p)^{n-r} \leqslant \tfrac{1}{40}.$$

The values of A and W can be found by evaluating the binomial probabilities or from National Bureau of Standards (1950). Alternatively, if the Poisson approximation is appropriate, they can be found by evaluating the Poisson probabilities or by consulting Pearson and Hartley (1966). So long as the number of defectives in a sample is less than W, it can be assumed that the process is in control. However one observation greater than or equal to A, or two consecutive results greater than or equal to W would mean that some action must be taken.

Control charts based on the number of defective items in a sample are sometimes called p-charts. However, with a complex product, it is possible that several defects may occur in the same item. Then the total number of defects in a sample can be calculated and plotted on a control chart which is sometimes called a c-chart (see, for example, Bowker and Lieberman, 1957).

12.9.1 *Tolerance limits and specification limits*

Control charts can also be used in a different type of situation where there is a zone of acceptable quality rather than a single target value. For example, the upper and lower limits for the design specifications for the width of a bolt may be 1·120 inches and 1·100 inches. Here a possible target value would be half-way between the specification limits but it would not matter if the over-all average measurement was somewhat different from this provided that most of the measurements still lay within the specification limits.

Suppose that a series of samples, size n, are taken and that a quantitative measurement is made on each item in the sample. An estimate of the population mean, μ, can be obtained by finding the over-all average of the sample means, which will be denoted by $\bar{\bar{x}}$. An estimate of the population standard deviation, σ, can be obtained from the average sampling range, $\bar{R}$, by the formula $\hat{\sigma} = \bar{R}/k_1$ (see Table 24). Tolerance limits for the actual measurements are usually given by $\mu \pm 3\sigma$, so that, if the observations are normally distributed, only 0·27 per cent of the observations will fall outside them. With the process under control, the tolerance limits can be estimated by $\bar{\bar{x}} \pm 3\bar{R}/k_1$. If these values do not fall inside the specification limits, then action should be taken to adjust the population mean or to reduce the population standard deviation. Note particularly that if the difference between the upper and lower specification limits is less than about 6σ, then the process is bound to produce some defective items even when it is under control. In such a case it may be necessary to revise the specification limits if it is impossible to reduce σ.

If the tolerance limits do fall inside the specification limits, then the process is operating at an acceptable level. Control charts for the sample mean and range can then be constructed as before except that the target value for the sample mean is replaced by $\bar{\bar{x}}$. Here we assume that the tolerance limits are not substantially smaller than the specification limits. If they are, it may be that 'too good' a product is being made and that production costs could be reduced by lowering standards somewhat. Alternatively it may be sensible to reduce the specification limits. If this is not done then it is possible for results to fall outside the control limits when the product is still well within the design specifications. This would go against the general rule that control charts should always be constructed so that it is practicable to investigate every point which falls outside the control limits. The reader is referred, for example, to Bowker and Lieberman (1957) and Rickmers and Todd (1967).

Example 2

The upper and lower specification limits for the width of a bolt are 1·120 inches and 1·100 inches. Fifteen samples of five measurements give the following readings for $\bar{x}$ and R. (Range values are multiplied by 10^3.) Set up control charts and see if the process is under control. (Note that in practice it is advisable to have at least twenty-five samples in order to set up control charts.)

Sample	$\bar{x}$	R	Sample	$\bar{x}$	R	Sample	$\bar{x}$	R
1	1·115	18	6	1·112	5	11	1·113	6
2	1·116	17	7	1·114	5	12	1·114	4
3	1·114	8	8	1·112	7	13	1·111	3
4	1·112	6	9	1·113	3	14	1·113	5
5	1·114	7	10	1·111	4	15	1·111	7

This gives $\bar{\bar{x}} = 1\cdot1130$ and $\bar{R} = 7\cdot0$. Action lines for the $\bar{x}$ chart are placed at

$$\bar{\bar{x}} \pm \frac{3\bar{R}}{k_1\sqrt{5}} = 1\cdot1130 \pm 0\cdot0040.$$

All the observed values of $\bar{x}$ are inside these control limits. The upper action line for the R-chart is placed at

$$\frac{k_3\bar{R}}{k_1} = 16\cdot4.$$

The values of R from samples 1 and 2 are greater than this and so the process was then out of control.

The values of $\bar{\bar{x}}$ and $\bar{R}$ were recomputed from samples 3–15. These were $\bar{\bar{x}} = 1\cdot1126$ and $\bar{R} = 5\cdot4$. The control limits for the $\bar{x}$ and R-charts

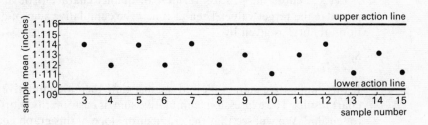

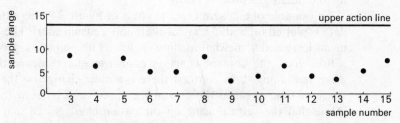

Figure 73   (a) $\bar{x}$-chart (b) R-chart

were also recomputed. The action lines for the $\bar{x}$-chart are placed at

$1{\cdot}1126 \pm 0{\cdot}0031$.

The upper action line for the R-chart is placed at $12{\cdot}7$. The values of $\bar{x}$ and R from samples 3 to 15 are plotted in Figure 73. Since none of the points fall outside the control limits, we can assume that the process is now under control.

An estimate of the tolerance limits is given by $\bar{\bar{x}} \pm 3\bar{R}/k_1 = 1{\cdot}1126 \pm 0{\cdot}0070$. These are inside the specification limits and so the production material will come up to the required standards provided that the process remains in control at the above levels.

12.10 Cusum charts

We have seen that control charts are a valuable aid in detecting changes in process performance. However the procedure is rather crude in that small changes in the process mean are often obscured by the residual variation. This is because each observation is independently compared with the control limits whereas it would appear sensible to combine successive results in some way (as in the rule of seven). It turns out that changes in the process mean are often easier to detect with the aid of *cusum charts*, which are a new and valuable device in statistical quality control.

Let x_p denote the process variable or quality characteristic at time p, and T the target value. Then at time t the cumulative sum of deviations about T is given by

$$S_t = \sum_{p \leqslant t} (x_p - T).$$

If this cumulative sum is plotted against time then a cumulative sum chart results. The expression 'cumulative sum' is customarily shortened to 'cusum'. We will see that the gradient or slope of this graph enables us to estimate the process mean.

An example of a cusum chart is given in Figure 74 where the same data is plotted on both a control chart and a cusum chart. The process mean increased somewhat at time t_1, but as no point lies outside the action lines, the control chart gives no definite evidence that the mean has changed. In contrast there is a clear change on the cusum chart at time t_1 since the cusum increases steadily from there on. [Note that the vertical scale on the cusum chart is half that on the control chart.]

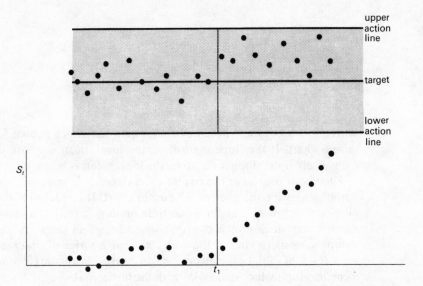

Figure 74 A control chart compared with a cusum chart

The calculation of the cusums is a very simple matter. We have

$$S_1 = x_1 - T,$$

$$S_2 = (x_1 - T) + (x_2 - T)$$
$$= S_1 + (x_2 - T).$$

In general

$$S_t = S_{t-1} + (x_t - T).$$

The next problem is to show that the local process mean depends upon the *slope* of the cusum graph. We will show that the slope of the line joining S_m to S_n measures the average difference from T of x_{m+1} to x_n.

The local process mean, $\bar{x}_\text{L}$, is given by

$$\bar{x}_\text{L} = \sum_{i=m+1}^{n} \frac{x_i}{n-(m+1)}$$

$$= T + \frac{\sum_{i=m+1}^{n} (x_i - T)}{n-(m+1)}$$

$$\bar{x}_L = T + \frac{S_n - S_m}{n - (m+1)}$$

$$= T + \frac{\text{(change in cusum)}}{\text{(number of observations)}}$$

$$= T + k \times \text{slope},$$

where k is a constant which depends upon the scales chosen for the cusum chart. If the slope is positive, the local mean is above target. Conversely if the slope is negative, the local mean is below target.

The visual impact of the cusum chart depends in part on choosing 'good' scales for the axes of the cusum chart. It is advisable to keep all slopes below 60°, at the same time making sure that a change in mean really does give a clearly visible change of slope. A suitable compromise is to choose the scales so that a series of observations, with residual variation σ, whose mean moves 2σ off target will give a cusum graph which makes 45° with the horizontal.

Occasionally there will be no obvious target value and then the cusums must be calculated with respect to some carefully chosen reference value. This reference value must be close to the average of the observations as it is much easier to spot changes in the mean when the slope of the cusum graph changes sign.

It may be possible to detect changes from one positive slope to a different positive slope but this is much more difficult. In any case if the reference value is less than most of the observations, then the cusum will increase rapidly and will quickly run off the graph paper.

12.10.1 Detection of changes

It is often possible to detect when a change in the process has occurred by visual inspection of the cusum chart. However we will describe two objective methods which have been proposed; the first by Barnard and the second by Ewan and Kemp. These two methods have been shown to be equivalent.

V-mask technique. A V-shaped mask is placed on the cusum chart at a distance d ahead of the latest observation. The angle between the limbs of the V-mask and the horizontal is denoted by θ. If all the previous observations lie inside the V-mask, as in Figure 75, then the process is assumed to be in control. Conversely if one or more observations lie outside the V-mask then the process mean is assumed to have changed.

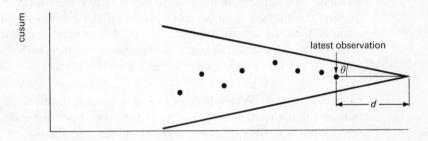

Figure 75   A V-mask

The properties of the V-mask depend on the choice of d and θ. If d and θ are 'large' then the V-mask will rarely indicate a change in process mean. Conversely if d and θ are 'small' then there will be lots of interruptions. We want to choose d and θ so as to detect any real changes quickly but in such a way that an interruption is unlikely if no real change has occurred.

The values of d and θ can be chosen by trying different V-masks on past data and selecting one which gives reliable results. Alternatively they can be obtained by considering average run lengths. For a particular value of the process mean the average run length is the average number of observations which are taken before the V-mask indicates that a change has occurred. The average run length should be high when the process is on target (to avoid false alarms) but low when the process is off target. These run lengths have been calculated by Monte Carlo methods for different values of d and θ and are tabulated in I.C.I. (1964). The calculations assume that the observations follow a normal distribution with variance σ^2 and that the scales have been standardized, as mentioned before, so that the horizontal distance between successive points is equal to 2σ on the vertical axis. Results are given for displacements of the process mean varying between zero and 3σ. For example, if $d = 2$ and $\tan \theta = 0.5$, the average run length is 140 when the process is on target but only three when the process is 2σ off target. The manufacturer should choose two such run lengths and hence find the appropriate V-mask.

Decision interval. This scheme depends on two quantities called the reference value and the decision interval. The reference value, R, is chosen midway between the target value and an unsatisfactory level. Every time a reading exceeds the reference value a cusum is started in which the calculations are made with respect to R. If the cusum returns to zero the process is under control. But if the process mean has changed then the cusum will increase and will eventually exceed the decision interval, at which point the process is assumed to be out of control. Note that no plotting is necessary for this scheme.

Corrective action. When the V-mask or decision interval scheme indicates that a change has occurred in the process performance, then some form of corrective action must be taken. The decision interval approach can be adapted to give an estimate of the current process mean whenever an 'action' signal occurs. The change in mean will correspond to a known change in the setting of one of the process input variables.

12.10.2 *Post-mortems*

Cusum charts can also be used to analyse data retrospectively. Instead of deducting some target value, we deduct the mean of the observations over the period in question. This means that the cusum starts and finishes at zero. The main point of interest is to determine if and when changes in the local mean occurred. There are several ways of doing this. The method we describe is suitable for computers.

First any freak values are removed from the data and replaced by the average of the two adjacent values. The residual variance may be estimated by

$$s^2 = \sum_{i=1}^{n-1} \frac{(x_{i+1} - x_i)^2}{2(n-1)} \quad \text{with } (n-1) \text{ d.f.}$$

Then Student's t-test is used to locate the turning points; that is the points at which the local mean changes.

This is done in the following way. Move along the series by connecting each observation with the last turning point (or initially with the first observation), and find the maximum absolute distance between this chord and the intervening cusum readings. The position of this largest difference is used to divide the intervening readings into two groups size n_1 and n_2 respectively. The means $\bar{x}_1$ and $\bar{x}_2$ of the two groups are calculated. If the process mean is unchanged throughout the interval then the standard error of the difference between $\bar{x}_1$ and $\bar{x}_2$ is

$$s \sqrt{\left(\frac{1}{n_1} + \frac{1}{n_2} \right)}.$$

The required test statistic is given by

$$t = \frac{\text{Difference between the means}}{\text{Standard error of the difference}}$$

$$= \frac{\bar{x}_1 - \bar{x}_2}{s \sqrt{\left(\frac{1}{n_1} + \frac{1}{n_2} \right)}}.$$

If the value is significantly large then we reject the null hypothesis that the process mean is unchanged and assume that we have found a turning point. However if the value is not significant, then we move to the next observation on the cusum chart and repeat the process. Turning points can also be found by moving backwards along the series. The two lists of turning points are amalgamated.

This type of repetitive analysis is ideal for a computer to perform. In addition, by suitably choosing the scales, the cusum chart can be printed out together with a Manhattan diagram of the local means, which, as the name implies, is a block diagram as illustrated in Figure 76.

12.10.3 *When to use cusum charts*

Given a series of observations which are sequentially arranged in time, it is usually more efficient to use a cusum chart rather than a control

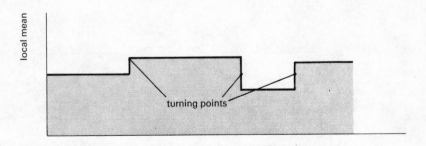

Figure 76   A Manhattan diagram

chart to detect changes in the local mean. However if we are only interested in changes which are 'large' compared with the residual variation, then a control chart may be adequate. In this case the control chart is preferred on grounds of simplicity as the non-statistician finds them easier to understand than cusum charts.

12.11 Feedback control

Quality control charts, such as Shewhart and cusum charts, are one type of method for controlling industrial processes. We will now consider the problem in more general terms. The uncontrolled system can be represented diagrammatically as follows:

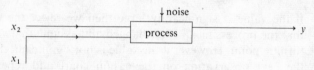

x_1 represents the input variables which can be controlled;
x_2 represents the input variables which cannot be controlled though it may be possible to measure them and
y represents the output variables.

In addition the output variables will be affected by a series of random disturbances, which is called *noise* by electrical engineers.

The purpose of control is to feed back the information gained from measuring y in order to adjust x_1 and so maximize (or minimize) some performance criterion (for example, cost, output or efficiency). Such a system is called a *feedback control system*. It can be represented diagrammatically as follows:

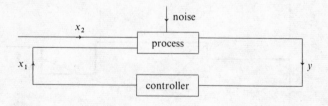

This is sometimes called a closed loop system for an obvious reason. The controller may be

(a) A man with control charts. This type of control is called *statistical quality control*.

(b) An analogue computer. This type of control is called *automatic control*.

(c) A digital computer. This type of control is called *on-line* or *off-line control* according as to whether or not the computer is connected directly to the process.

Statisticians in the past have dealt mainly with the problem of detecting changes in process performance (usually with control charts) and have paid little attention to the problem of applying corrections continuously to the input variables, as a result of measurements on the output variables. However a lot of theoretical work on control problems has been carried out particularly by electrical engineers, and there is a well-developed theory of *deterministic* control which has proved very useful. The study of *stochastic* control, in which the system is affected by random disturbances, was pioneered by Wiener among others. Much of the published work depends on an exact knowledge of the statistical properties of the noise and sometimes also those of the signal. In addition it may also be necessary to have exact knowledge of the linear differential equations which govern the behaviour of the system. The noise is usually assumed to be a stationary purely random Gaussian (i.e. normal) time-series. This approach has played an important role in the mathematical development of control theory and in such applications as servomechanism design. But in many situations the theory may not be very helpful, partly because the differential equations may be unknown and partly because the noise may have unknown statistical properties or may be non-stationary. Recent developments include studies of nonlinear control and of systems affected by non-stationary noise. The reader is referred to Wiener (1949), Kalman (1960), Florentin (1961) and Fel'dbaum (1966).

An alternative more practical method of control, has been suggested by Box and Jenkins (1962). They point out that prediction and control are closely inter-related so that if it is possible to predict how a process will behave then it should also be possible to control it. For example, if a prediction is made that the response variable will be a certain distance off target then the input variables can be changed by the appropriate amount to bring the response variable back on target. Box and Jenkins then go on to consider the problem of prediction in detail and describe a method which is a generalization of a well known forecasting method called *exponential smoothing*.

Given a series of reading $x_1, x_2, \ldots, x_p$ we want to predict x_{p+1} or more generally x_{p+q}. Denote the prediction made at time p of the reading q steps ahead by $\hat{x}(p, q)$. A general class of estimates is given by

$$\hat{x}(p, 1) = W_0 x_p + W_1 x_{p-1} + \ldots + W_k x_{p-k} + \ldots,$$

where W_i are weights which are applied to the previous observations. One set of weights, which has given consistently good results for series with no obvious trend, is the set of geometric weights given by

$$W_k = \alpha(1-\alpha)^k \qquad (0 < \alpha < 1).$$

These weights are intuitively appealing since they sum to one and also decrease as the corresponding readings go further into the past. Moreover they give rise to the simple recursive formula

$$\hat{x}(p, 1) = \alpha x_p + (1-\alpha)\hat{x}(p-1, 1).$$

This can be rewritten as

$$\hat{x}(p, 1) = \alpha e_p + \hat{x}(p-1, 1),$$

where $\quad e_p = x_p - \hat{x}(p-1, 1)$

$\qquad\qquad = $ error made in prediction at time p.

This form of prediction is called exponential smoothing. However if there is a trend or offset in the data then the predictions may be systematically above or below target. Thus the above equation was generalized to give

$$\hat{x}(p, 1) = \hat{x}(p-1, 1) + \gamma_{-1}\Delta e_p + \gamma_0 e_p + \gamma_1 \sum_{j \leqslant p} e_j,$$

where $\quad \gamma_0 = \alpha$

and $\quad \Delta e_p = e_p - e_{p-1}.$

The three terms in this prediction formula are referred to as the first difference, proportional and cumulative terms, or, in engineering terms, the derivative, proportional and integral terms. The next problem is to estimate the three unknown parameters, γ_{-1}, γ_0 and γ_1. This can be done by choosing them so that the sum of squared errors, made in predicting the observations already known, is minimized. Details of this computation, which may be done on a computer, are given in Box and Jenkins (1962).

When a prediction of the output has been made, the next problem is to calculate the necessary adjustments to the input variables which will keep the process on target. The problem is complicated by the fact that adjustments may not become effective immediately. If a step change is made to an input variable, some typical changes in the output are as shown in Figure 77.

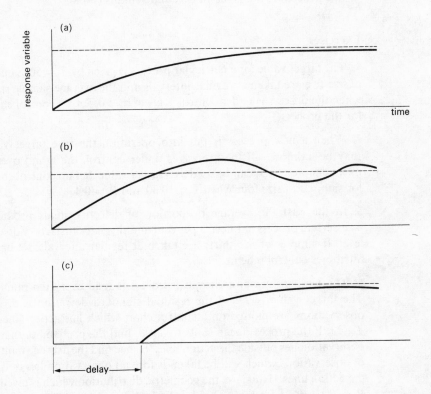

Figure 77   (a) Exponential response (b) Oscillatory response
(c) Delayed exponential

One model for describing the dynamics of the system is given in Box and Jenkins (1968). This model depends on several parameters. These are defined as follows. Firstly if a change of b units in the input variable produces an eventual change of a units in the output, then the *gain* of the system is defined as $g = a/b$. Secondly if a change in the input variable does not produce any change in the output for d units of time, then d is called the *delay* or *dead time*, and lastly if the response is exponential, the rate at which it converges depends on a parameter denoted by δ_1. Given a series of input and output readings it is possible to estimate these parameters by the method of least squares and hence compute the optimal adjustments to the input variable. Full details of this procedure are given in Box and Jenkins (1968).

Exercises

1. The target value for a particular process is given by $T = 50$. Samples of size five are taken at regular intervals and the average sample range when the process is under control is given by 5·0. Set up control charts for the process.

2. When a new process is put into operation the true target value may be unknown. If the process is under control, the grand over-all mean denoted by $\bar{\bar{x}}$, can be used as a target value. Set up control charts for samples of size four when $\bar{\bar{x}} = 10\!\cdot\!80$ and $\bar{R} = 0\!\cdot\!46$.

3. In the past the average proportion of defective items produced by a certain process has been 3 per cent and this is a satisfactory quality level. If samples of size thirty are taken at regular intervals, set up an attributes control scheme.

4. The temperature of a chemical process is read every ten minutes. The target value is T and the residual standard deviation is σ. The observations are plotted on a control chart which has action lines at $T \pm 3\sigma$. If the process mean shifts to $T + \sigma$, find the probability that an observation lies outside the action lines. Hence find the average number of observations which will be taken before an observation lies outside the action lines. [Hint: use the geometric distribution which is given by $P_r = \theta(1-\theta)^{r-1}$ for $r = 1, 2, \ldots$ and which has mean $1/\theta$.]

5. Warning lines are placed at $T \pm 2\sigma$ on the control chart discussed in question 4. Find the probability that two points in succession lie outside the warning lines when the process is on target. Also calculate this probability if the process mean shifts to $T + \sigma$.

6. The specifications for a certain quality characteristic are $15\!\cdot\!0 \pm 6\!\cdot\!0$ (in coded values). Fifteen samples of four readings gave the following values for $\bar{x}$ and R.

Sample	$\bar{x}$	R	Sample	$\bar{x}$	R	Sample	$\bar{x}$	R
1	16·1	3·0	6	15·7	2·7	11	15·3	13·8
2	15·2	2·1	7	15·2	2·3	12	17·8	14·2
3	14·2	5·6	8	15·0	3·8	13	15·9	4·8
4	13·9	2·4	9	16·5	5·0	14	14·6	5·0
5	15·4	4·1	10	14·9	2·9	15	15·2	2·2

Use the results from all fifteen samples to set up $\bar{x}$ and R control charts. Hence decide if the process is in control. If not, remove the doubtful samples, and recompute values for $\bar{x}$ and R, for use in succeeding samples. Estimate the tolerance limits and see if the process will meet the required specifications.

References

BARNARD, G. A. (1946), 'Sequential tests in industrial statistics', supplement to the *Journal of the Royal Statistical Society*, series B, vol. 8, pp. 1–21.

BOWKER, A. H., and LIEBERMAN, G. J. (1957), *Engineering Statistics*, Prentice-Hall.

BOX, G. E. P., and JENKINS, G. M. (1962), 'Some statistical aspects of adaptive optimization and control', *Journal of the Royal Statistical Society*, series B, vol. 24, pp. 297–343.

BOX, G. E. P., and JENKINS, G. M. (1968), 'Some recent advances in forecasting and control. Part 1', *Applied Statistics*, vol. 17, pp. 91–109.

COLUMBIA UNIVERSITY STATISTICAL RESEARCH GROUP (1945), *Sequential Analysis of Statistical Data: Applications*, Columbia University Press.

COLUMBIA UNIVERSITY STATISTICAL RESEARCH GROUP (1948), *Sampling Inspection*, McGraw-Hill.

DODGE, H. F., and ROMIG, H. G. (1944), *Sampling Inspection Tables*, Wiley.

DUNCAN, A. J. (1965), *Quality Control and Industrial Statistics*, Irwin, 3rd edn.

FEL'DBAUM, A. A. (1966), *Optimal Control Systems*, Academic Press.

FLORENTIN, J. J. (1961), 'Optimal control of continuous time Markov stochastic systems', *Journal of Electronic Control*, vol. 10, pp. 473–82.

I.C.I. (1964), *Cumulative Sum Techniques*, I.C.I. Monograph, Oliver & Boyd.

KALMAN, R. E. (1960), 'A new approach to linear filtering and prediction problems', *Transactions of the A.S.M.E.: Series D: Journal of Basic Engineering*, vol. 82, pp. 35–45.

NATIONAL BUREAU OF STANDARDS (1950), *Tables of the Binomial Probability Distribution*, U.S. Government Printing Office.

PEARSON, E. S. (1960), *Application of Statistical Methods to Industrial Standardization and Quality Control*: B.S. 600, British Standards Institution, reprinted.

PEARSON, E. S., and HARTLEY, H. O. (1966), *Biometrika Tables for Statisticians*, Cambridge University Press, 3rd edn.

RICKMERS, A. D., and TODD, H. N. (1967), *Statistics*, McGraw-Hill.

SHEWHART, W. A. (1931), *Economic Control of Quality of Manufactured Product*, Van Nostrand.

U.S. DEPARTMENT OF DEFENSE (1963), *Military Standard 105D*, U.S. Government Printing Office.

WALD, A. (1947), *Sequential Analysis*, Wiley.

WIENER, N. (1949), *Extrapolation, Interpolation and Smoothing of Stationary Time-Series*, Wiley.

Chapter 13
Life testing

13.1 Problems in measuring reliability

The topic of *reliability* is of major importance in all manufacturing industries, whether the product concerned is a washing machine, a camera, a rocket engine, a car, an electron tube or any other durable consumer product. The reliability of a product is a measure of its quality and has a variety of meanings depending on the particular situation. For example, it may be the probability that a device will function successfully for a certain period of time, or it may be the expected fraction of time that a system functions correctly. The different definitions of reliability are discussed in Barlow and Proschan (1965), Chapter 1.

The problem of measuring reliability depends on a number of considerations. Firstly there are many products, like cars, which are repaired when they break down. Here we are not only interested in the time before the first failure but also in the times between subsequent failures. Secondly there are products which are not repaired (or cannot be repaired) and then the first failure time is also the life time of the product. In this introductory chapter we will only be concerned with some of the problems involved in finding a mathematical model to describe the distribution of the first failure times. However this is applicable to both the above types of product as the car manufacturer, for example, is more concerned with the first failure time than with subsequent breakdowns because of the guarantee which is given with the product.

One method of measuring the reliability of a product is to test a batch of items over an extended period of time and to note the failure times. This process is called *life-testing*. Thus in Example 3, Chapter 1, we have the failure times of a batch of refrigerator motors. The life-testing programme should provide answers to the following questions.

(1) What is the mean life of the product?

(2) If a guarantee is issued with each item, what proportion of the manufactured items will fail before the guarantee expires?

(3) What mathematical model will satisfactorily describe the distribution of failure times?

(4) If there is a change in the manufacturing process, how does the distribution of failure times alter?

Various type of life tests are available.

1. *Non-replacement*. N items are tested simultaneously under identical conditions, but an item is not replaced when it fails. The test continues either until all the items have failed, or until a specified time has elapsed (time-truncated), or until a specified number of failures have occurred (sample-truncated).

2. *Replacement*. N items are tested simultaneously under identical conditions. When an item fails, it is repaired or replaced. The test continues until a stopping rule is satisfied or it may continue indefinitely.

3. *Sequential*. When a new product is being tested a sequential operation can be useful. If a lot of failures occur early on and the product is obviously unsatisfactory, there may be no point in continuing the test. Conversely, if very few failures have occurred after a long period of testing, the product is obviously satisfactory and there may also be no point in continuing the test. Thus after each failure there are three possibilities.

(a) Product life time has been proved satisfactory; stop testing.
(b) Product life time has been proved unsatisfactory; stop testing.
(c) Not enough information is available; continue testing.

Before setting up the test programme the engineer must answer the following questions.

(1) How many items should be tested?

(2) How long should the test last?

(3) What constitutes a failure? Many products do not fail suddenly but deteriorate gradually, and it is essential to decide beforehand the decrease in performance which constitutes a failure. For commercial products like a washing machine, the failure point could be at the time when a service call would be expected.

(4) What precautions should be taken to ensure that the items chosen for the test are representative?

(5) Will the test be carried out under normal operating conditions?

It is obviously desirable, if possible, to carry out the tests under the same conditions as would be met in ordinary use. Unfortunately this would often require an unrealistic waiting time before useful results become available. For example, the guarantee on a refrigerator is usually one or two years and so tests under normal conditions would last several years.

There are two ways of accelerating life testing. The first method, called *compressed-time testing*, requires that the product should be used more intensively than usual but without changing the stress levels to which the product is subjected. For example, a washing machine is often used just once a week. By running the machine several times a day, the machine can be made to do as much work in a few weeks as it would normally do in a year or two, so that useful results would become available much quicker than would otherwise be the case.

The second method of accelerated life-testing is called *advanced-stress testing* and consists of subjecting the product to higher stress levels than would normally apply. Such tests will induce failures in a much shorter time. For example, a refrigerator is normally in use all the time so that it cannot be subjected to compressed-time testing. When failures do occur they are usually caused by the refrigerator motor. By running the motor at higher speeds than usual it is found that failures occur within a few weeks. Past experience may suggest that say one week under advanced stress conditions is equivalent to one year under normal operating conditions, and this enables information to be derived from the data obtained under advanced stress conditions.

Data obtained from accelerated life testing is of course much less reliable than that obtained under normal conditions and should always be treated with suspicion. However it may often be the only way of getting information in the time available.

13.2 The mathematical distribution of failure times

Since the failure time of an item may be any positive number, the distribution of failure times is continuous. Thus it can be described either by its probability density function, $f(t)$, or by its cumulative distribution function, $F(t)$. Then the probability that an item will fail in the time interval from t to $t + \Delta t$ is given by $f(t)\Delta t$. Also the probability that an item will fail at any time up to t is given by

$$F(t) = \int_0^t f(t)\, dt.$$

It is also convenient to introduce a function $R(t)$ called the *reliability function* which gives the probability that an item will survive until time t. Thus we have

$$R(t) = 1 - F(t).$$

After testing a sample of items, we will have a series of observed failure times denoted by

$$0 \leqslant t_1 \leqslant t_2 \ldots \leqslant t_n.$$

Thus it appears that our problem is to estimate the related functions $f(t)$, $F(t)$ and $R(t)$ from the observed failure times. However if the type of mathematical distribution is unknown it is often expedient to estimate another function called the *conditional failure rate function* (or hazard function or instantaneous failure rate function) which is given by

$$Z(t) = \frac{f(t)}{R(t)}.$$

Then the probability that an item will fail in the interval from t to $t + \Delta t$, *given that it has survived until time t*, is given by $Z(t)\Delta t$. It is most important to realize that this is a conditional probability and that $Z(t)$ is always larger than $f(t)$, because $R(t)$ is always less than 1. For example, the proportion of cars which fail between fifteen and sixteen years is very small because few cars last as long as fifteen years. But of the cars which do last fifteen years a substantial proportion will fail in the following year. In fact the function $Z(t)$ describes the distribution of failure times just as completely as $f(t)$ or $F(t)$, and is perhaps the most basic measure of reliability.

It is useful to establish a direct relation between $Z(t)$ and $R(t)$. This is done in the following way. The functions $f(t)$ and $F(t)$ are related by the equation

$$f(t) = \frac{dF(t)}{dt} \qquad \text{(see section 5.1)}$$

and $R(t) = 1 - F(t)$,

$$f(t) = -\frac{dR(t)}{dt}.$$

Thus $Z(t) = \frac{f(t)}{R(t)}$

$$= \frac{-[dR(t)]/dt}{R(t)}$$

$$= -\frac{d \ln R(t)}{dt}.$$

Conversely $\quad R(t) = \exp\left[-\int_0^t Z(t)\, dt\right].$ $\qquad$ **13.1**

13.3 The exponential distribution

This distribution was introduced in section 5.6 and is the simplest theoretical distribution for describing failure times.

Suppose a product is such that its age has no effect on its probability of failing, so that, if it has survived until time t, the probability that it will fail in the interval t to $(t + \Delta t)$ is given by $\lambda \Delta t$, where λ is a constant for all t. In other words the conditional failure rate, $Z(t)$, of the product is a constant for $t > 0$.

$$Z(t) = \lambda.$$

Then $\quad R(t) = \exp\left[-\int_0^t Z(t)\, dt\right]$

$$= \exp[-\lambda t],$$

$$F(t) = 1 - \exp[-\lambda t],$$

$$f(t) = \frac{dF(t)}{dt}$$

$$= \lambda \exp[-\lambda t].$$

We notice that $f(t)$ is the probability density function of the exponential distribution.

The exponential model states that chance alone dictates when a failure occurs and that the product's age does not affect the issue. There are a few products of this type such as electron tubes and electric fuses. In addition the exponential distribution describes the times between failures of a complex piece of equipment (see Barlow and Proschan, 1965, Chapter 2). However the reader is warned that many products deteriorate with age and then the exponential model will not apply.

The mean failure time for the exponential distribution is given by

$$E(t) = \int_0^\infty t\lambda\, e^{-\lambda t}\, dt = \frac{1}{\lambda}.$$

Estimation. If the exponential model is thought to be appropriate, the next problem is to use the sample data to estimate λ, or its reciprocal $\theta = 1/\lambda$, which is the mean life. If all the items in the sample are run to destruction without replacement then the mean sample failure time is the best estimate of θ. However if the test programme is time-truncated or sample-truncated, or if a replacement programme is used, then an alternative method of estimation is required.

If N items are tested without replacement until n failures have occurred and if t_i denotes the ith failure time, then an estimate of θ is given by

$$\hat{\theta} = \frac{t_1 + t_2 + \ldots + t_n + (N-n)t_n}{n}.$$

If N items are tested with replacement until n failures have occurred, we will denote the ith failure time by t_i, whether it applies to an original or a repaired item. Each of these times is measured from the beginning of the experiment. Then an estimate of θ is given by

$$\hat{\theta} = \frac{Nt_n}{n}.$$

The reader is referred to Epstein (1954) for a discussion of these results and for a discussion of the problem of testing $H_0 : \theta = \theta_0$ against the alternative hypothesis $H_1 : \theta < \theta_0$.

13.4 Estimating the conditional failure rate

The conditional failure rate, $Z(t)$, is one of the basic measures of the reliability of a product. In the previous section we considered the case where $Z(t)$ is constant and the failure times follow an exponential distribution. But many situations exist in which $Z(t)$ is not constant and, in the absence of prior information, it is always a good idea to estimate the shape of the function. This will provide information about the way the product fails and give insight into the type of mathematical model which will be appropriate. We will describe a method of doing this when a non-replacement programme is used.

N items are tested simultaneously. Let t_i denote the time at which the ith failure occurs. Let Δt_i denote the time between the ith failure and the $(i+1)$th failure and let Δt_0 denote the time before the first failure. Now $Z(t)\Delta t$ is the conditional probability of failure in the interval t to $t+\Delta t$, given that the item has survived until time t. Thus $Z(t_i)\Delta t_i$ can be estimated by dividing the observed number of failures in the interval Δt_i (which by definition is one) by the number of items which last longer than t_i (which is $N-i$). Thus the estimate of $Z(t_i)$ is given by $1/(N-i)\Delta t_i$.

Example 1

Estimates of $Z(t_i)$ were obtained for the data of Example 3, Chapter 1, and the results are given below.

i	t_i	Δt_i	$(N-i)\Delta t_i$	$\hat{Z}(t_i)$
0	0	104·3	2086·0	0·0005
1	104·3	54·4	1033·6	0·001
2	158·7	35·0	630·0	0·002
3	193·7	7·6	129·2	0·008
4	201·3	4·9	78·4	0·013
5	206·2	21·6	324·0	0·003
6	227·8	21·3	298·2	0·003
7	249·1	58·7	763·1	0·001
8	307·8	3·7	44·4	0·022
9	311·5	18·1	199·1	0·005
10	329·6	28·9	289·0	0·003
11	358·5	5·8	52·2	0·019
12	364·3	6·1	48·8	0·020
13	370·4	10·1	70·7	0·014
14	380·5	14·1	84·6	0·012
15	394·6	31·6	158·0	0·006
16	426·2	7·9	31·6	0·032
17	434·1	118·5	355·5	0·003
18	552·6	41·4	82·8	0·012
19	594·0	97·5	97·5	0·010
20	691·5			

The estimates of $Z(t_i)$ range from less than 0·001 to 0·032 and the values are plotted in Figure 78. It appears that $Z(t)$ is increasing, in which case the exponential model will not apply. We will describe a more suitable model in the next section.

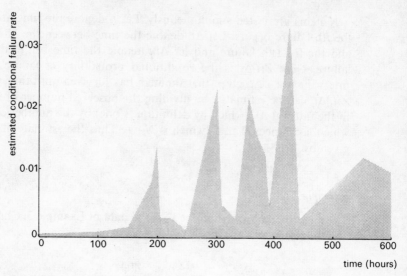

Figure 78

An increase in $Z(t)$ is a common phenomenon and means that the product is more likely to fail the older it gets. Another common type of conditional failure rate function is depicted in Figure 79. Teething troubles due to faulty manufacture are quite likely to occur so that $Z(t)$ is relatively large for low values of t. Then there is a period when the product is likely to be trouble free, so that $Z(t)$ is relatively low. Finally, as the product ages, more failures occur and $Z(t)$ rises steadily.

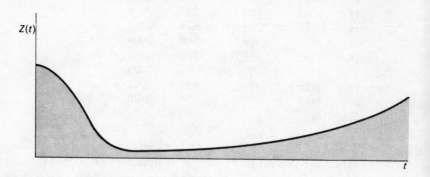

Figure 79   A common type of conditional failure rate

It is unrealistic to expect one model to describe all possible situations. However a model based on the *Weibull* distribution is widely applicable and will be described in the next section.

13.5   The Weibull distribution

The Weibull (1951) distribution has been used successfully to describe among other things, fatigue failures, vacuum tube failures and ball bearing failures, and is probably the most popular model for describing failure times. The conditional failure rate for the Weibull model is given by

$$Z(t) = m\lambda t^{m-1}.$$

This function depends on two parameters, m and λ, which are called the Weibull shape parameter and the Weibull scale parameter respectively. For the special case when $m = 1$, we find $Z(t) = \lambda$, so the exponential model is a special case of the Weibull model. Various types of conditional failure rates are shown in Figure 80. For values of m greater than one, $Z(t)$ increases with t. Thus the Weibull model may describe the data which was analysed in Example 1. However the Weibull model will not give a function, similar to that depicted in Figure 79, for any value of m and so the model will not apply in this case.

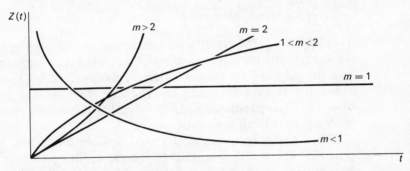

Figure 80   Some Weibull conditional failure rates

Using formula **13.1** we can find the reliability function of the Weibull model,

$$R(t) = \exp[-\lambda t^m].$$

Hence $F(t) = 1 - \exp[-\lambda t^m]$

and $f(t) = m\lambda t^{m-1}\exp[-\lambda t^m].$

The last two functions are the cumulative distribution function and the probability density function of the Weibull distribution.

If the Weibull distribution is thought to be appropriate, then both Weibull parameters must be estimated from the data. Efficient methods of estimation are available in the literature (see Buckland, 1964, and Cohen, 1965), and these should be used if such calculations are to be made regularly. Here we describe a simple graphical method of estimation which is particularly useful in the early stages of an investigation as it can be also used to see if the Weibull model is appropriate.

The Weibull reliability function is given by

$$R(t) = \exp[-\lambda t^m],$$

giving $\ln R(t) = -\lambda t^m$

$$\ln[-\ln R(t)] = m \ln t + \ln \lambda.$$

N items are tested without replacement. Let t_i denote the ith failure time. Then it can be shown that an unbiased estimate of $F(t_i)$ is given by $i/(N+1)$. Thus an unbiased estimate of $R(t_i)$ is given by $(N+1-i)/(N+1)$. But

$$-\ln\left(\frac{N+1-i}{N+1}\right) = \ln\left(\frac{N+1}{N+1-i}\right).$$

Thus if $\ln[(N+1)/(N+1-i)]$ is plotted against t_i on log–log graph paper (on which both scales are logarithmic), the points will lie approximately on a straight line if the Weibull model is appropriate. It will usually be sufficient to fit a straight line by eye. The slope of the line is an estimate of m. The other parameter λ can be estimated at the value of t where $\ln[(N+1)/(N+1-i)]$ is equal to one.

Before giving an example of this method of estimation we will generalize the model somewhat to include a third parameter. This quantity specifies the minimum life time of the product and is sometimes called the Weibull location parameter. It will be denoted by L. The conditional failure rate of the three-parameter Weibull model is given by

$$Z(t) = \begin{cases} m\lambda(t-L)^{m-1} & L < t < \infty, \\ 0 & \text{otherwise.} \end{cases}$$

Thus no failures occur before L. Then we have

$$R(t) = \exp[-\lambda(t-L)^m]$$
$$F(t) = 1 - \exp[-\lambda(t-L)^m] \left.\vphantom{\begin{array}{c}a\\b\\c\end{array}}\right\} \quad L < t < \infty.$$
$$f(t) = m\lambda(t-L)^{m-1}\exp[-\lambda(t-L)^m]$$

A crude estimate of L can be obtained from the smallest failure time. This is subtracted from all the failure times and then the remaining observations can be analysed as before, in order to obtain estimates of m and λ. This method will be illustrated with the data of Example 3, Chapter 1, although the introduction of the location parameter, L, is probably not really necessary in this case.

Example 2

The smallest failure time is 104·3 hours, so $\hat{L} = 104\cdot3$. This is subtracted from all the other observations and the analysis is carried out on the remaining nineteen observations. Thus we take $N = 19$ and then t_i will be the $(i-1)$th failure after $\hat{L}$. The values of $\ln[20/(20-(i-1))]$ are given below and the resulting values are plotted on log–log paper in Figure 81.

i	$t_i - 104\cdot3$	$\dfrac{20}{21-i}$	$\ln\left(\dfrac{20}{21-i}\right)$
1	0		
2	54·4	1·05	0·049
3	89·4	1·11	0·104
4	97·0	1·17	0·156
5	101·9	1·25	0·222
6	123·5	1·33	0·285
7	144·8	1·43	0·358
8	203·5	1·54	0·432
9	207·2	1·67	0·513
10	225·3	1·82	0·600
11	254·2	2·00	0·691
12	260·0	2·22	0·798
13	266·1	2·50	0·912
14	276·2	2·86	1·05
15	290·3	3·33	1·20
16	321·9	4·00	1.38
17	329·8	5·00	1·61
18	448·3	6·67	1·90
19	489·7	10·00	2·30
20	587·2	20·00	3·00

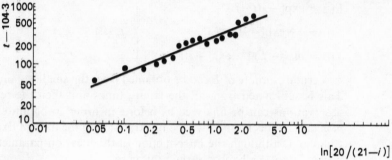

Figure 81

The points lie approximately on a straight line, indicating that the Weibull model really is appropriate. A straight line is fitted to the data by eye. From this we find

$\hat{m} = 1 \cdot 72 = $ (slope of the line)

and when $\ln\left(\dfrac{20}{21-i}\right) = 1$,

the corresponding value of t enables us to estimate λ by

$\hat{m}\ln(t - \hat{L}) = -\ln \hat{\lambda}$.

From the graph we see $t - \hat{L} = 290$ hours when $\ln[20/(21-i)] = 1$.

Hence $\hat{\lambda} = 290^{-1 \cdot 72}$

$\qquad = 0 \cdot 000059$.

The exponential and Weibull distributions are only two of the many possible types of model for describing the failure times of manufactured products. Other distributions which have been suggested include the gamma and log–normal distributions. The choice of a suitable distribution is by no means automatic. The reader is referred to Barlow and Proschan (1965) and Buckland (1964) for a general discussion of the problem.

Appendix B **Statistical tables**

Table 1 Areas under the normal curve

	·00	·01	·02	·03	·04	·05	·06	·07	·08	·09
0·0	·5000	·5040	·5080	·5120	·5160	·5199	·5239	·5279	·5319	·5359
·1	·5398	·5438	·5478	·5517	·5557	·5596	·5636	·5675	·5714	·5753
·2	·5793	·5832	·5871	·5910	·5948	·5987	·6026	·6064	·6103	·6141
·3	·6179	·6217	·6255	·6293	·6331	·6368	·6406	·6443	·6480	·6517
·4	·6554	·6591	·6628	·6664	·6700	·6736	·6772	·6808	·6844	·6879
·5	·6915	·6950	·6985	·7019	·7054	·7088	·7123	·7157	·7190	·7224
·6	·7257	·7291	·7324	·7357	·7389	·7422	·7454	·7486	·7517	·7549
·7	·7580	·7611	·7642	·7673	·7703	·7734	·7764	·7793	·7823	·7852
·8	·7881	·7910	·7939	·7967	·7995	·8023	·8051	·8078	·8106	·8133
·9	·8159	·8186	·8212	·8238	·8264	·8289	·8315	·8340	·8365	·8389
1·0	·8413	·8438	·8461	·8485	·8508	·8531	·8554	·8577	·8599	·8621
1·1	·8643	·8665	·8686	·8708	·8729	·8749	·8770	·8790	·8810	·8830
1·2	·8849	·8869	·8888	·8906	·8925	·8943	·8962	·8980	·8997	·9015
1·3	·9032	·9049	·9066	·9082	·9099	·9115	·9131	·9147	·9162	·9177
1·4	·9192	·9207	·9222	·9236	·9251	·9265	·9279	·9292	·9306	·9319
1·5	·9332	·9345	·9357	·9370	·9382	·9394	·9406	·9418	·9429	·9441
1·6	·9452	·9463	·9474	·9484	·9495	·9505	·9515	·9525	·9535	·9545
1·7	·9554	·9564	·9573	·9582	·9591	·9599	·9608	·9616	·9625	·9633
1·8	·9641	·9648	·9656	·9664	·9671	·9678	·9686	·9693	·9699	·9706
1·9	·9713	·9719	·9726	·9732	·9738	·9744	·9750	·9756	·9761	·9767
2·0	·9772	·9778	·9783	·9788	·9793	·9798	·9803	·9808	·9812	·9817
2·1	·9821	·9826	·9830	·9834	·9838	·9842	·9846	·9850	·9854	·9857
2·2	·9861	·9864	·9868	·9871	·9875	·9878	·9881	·9884	·9887	·9890
2·3	·9893	·9896	·9898	·9901	·9904	·9906	·9909	·9911	·9913	·9916
2·4	·9918	·9920	·9922	·9924	·9927	·9929	·9930	·9932	·9934	·9936
2·5	·9938	·9940	·9941	·9943	·9945	·9946	·9948	·9949	·9951	·9952
2·6	·9953	·9955	·9956	·9957	·9959	·9960	·9961	·9962	·9963	·9964
2·7	·9965	·9966	·9967	·9968	·9969	·9970	·9971	·9972	·9973	·9974
2·8	·9974	·9975	·9976	·9977	·9977	·9978	·9979	·9979	·9980	·9981
2·9	·9981	·9982	·9982	·9983	·9984	·9984	·9985	·9985	·9986	·9986
3·0	·9986	·9987	·9987	·9988	·9988	·9989	·9989	·9989	·9990	·9990
3·1	·9990	·9991	·9991	·9991	·9992	·9992	·9992	·9992	·9993	·9993
3·2	·9993	·9993	·9994	·9994	·9994	·9994	·9994	·9995	·9995	·9995
3·3	·9995	·9995	·9995	·9996	·9996	·9996	·9996	·9996	·9996	·9996
3·4	·9997	·9997	·9997	·9997	·9997	·9997	·9997	·9997	·9997	·9998
3·5	·9998	·9998	·9998	·9998	·9998	·9998	·9998	·9998	·9998	·9998
3·6	·9998	·9998	·9998	·9999	·9999	·9999	·9999	·9999	·9999	·9999

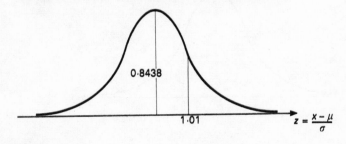

$$0·8438$$

$$1·01$$

$$z = \frac{x - \mu}{\sigma}$$

Table 2 Percentage points of Student's *t*-distribution

α	·01	·05	·025	·01	·005	·001
ν						
1	3·078	6·314	12·706	31·821	63·657	318·310
2	1·886	2·920	4·303	6·965	9·925	22·327
3	1·638	2·353	3·182	4·541	5·841	10·215
4	1·533	2·132	2·776	3·747	4·604	7·173
5	1·476	2·015	2·571	3·365	4·032	5·893
6	1·440	1·943	2·447	3·143	3·707	5·208
7	1·415	1·895	2·365	2·998	3·499	4·785
8	1·397	1·860	2·306	2·896	3·355	4·501
9	1·383	1·833	2·262	2·821	3·250	4·297
10	1·372	1·812	2·228	2·764	3·169	4·144
11	1·363	1·796	2·201	2·718	3·106	4·025
12	1·356	1·782	2·179	2·681	3·055	3·930
13	1·350	1·771	2·160	2·650	3·012	3·852
14	1·345	1·761	2·145	2·624	2·977	3·787
15	1·341	1·753	2·131	2·602	2·947	3·733
16	1·337	1·746	2·120	2·583	2·921	3·686
17	1·333	1·740	2·110	2·567	2·898	3·646
18	1·330	1·734	2·101	2·552	2·878	3·610
19	1·328	1·729	2·093	2·539	2·861	3·579
20	1·325	1·725	2·086	2·528	2·845	3·552
21	1·323	1·721	2·080	2·518	2·831	3·527
22	1·321	1·717	2·074	2·508	2·819	3·505
23	1·319	1·714	2·069	2·500	2·807	3·485
24	1·318	1·711	2·064	2·492	2·797	3·467
25	1·316	1·708	2·060	2·485	2·787	3·450
26	1·315	1·706	2·056	2·479	2·779	3·435
27	1·314	1·703	2·052	2·473	2·771	3·421
28	1·313	1·701	2·048	2·467	2·763	3·408
29	1·311	1·699	2·045	2·462	2·756	3·396
30	1·310	1·697	2·042	2·457	2·750	3·385
40	1·303	1·684	2·021	2·423	2·704	3·307
60	1·296	1·671	2·000	2·390	2·660	3·232
120	1·289	1·658	1·980	2·358	2·617	3·160
∞	1·282	1·645	1·960	2·326	2·576	3·090

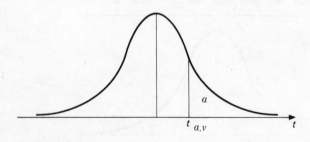

Table 3 Percentage points of the χ^2 distribution

α	·995	·99	·975	·95	·50	·20	·10	·05	·025	·01	·005
ν											
1	0·000	0·0002	0·001	0·0039	0·45	1·64	2·71	3·84	5·02	6·63	7·88
2	0·010	0·020	0·051	0·103	1·39	3·22	4·61	5·99	7·38	9·21	10·60
3	0·072	0·115	0·216	0·352	2·37	4·64	6·25	7·81	9·35	11·34	12·84
4	0·207	0·30	0·484	0·71	3·36	5·99	7·78	9·49	11·14	13·28	14·86
5	0·412	0·55	0·831	1·15	4·35	7·29	9·24	11·07	12·83	15·09	16·75
6	0·676	0·87	1·24	1·64	5·35	8·56	10·64	12·59	14·45	16·81	18·55
7	0·989	1·24	1·69	2·17	6·35	9·80	12·02	14·07	16·01	18·48	20·28
8	1·34	1·65	2·18	2·73	7·34	11·03	13·36	15·51	17·53	20·09	21·95
9	0·73	2·09	2·70	3·33	8·34	12·24	14·68	16·92	19·02	21·67	23·59
10	2·16	2·56	3·25	3·94	9·34	13·44	15·99	18·31	20·48	23·21	25·19
11	2·60	3·05	3·82	4·57	10·34	14·63	17·28	19·68	21·92	24·72	26·76
12	3·07	3·57	4·40	5·23	11·34	15·81	18·55	21·03	23·34	26·22	28·30
13	3·57	4·11	5·01	5·89	12·34	16·98	19·81	22·36	24·74	27·69	29·82
14	4·07	4·66	5·63	6·57	13·34	18·15	21·06	23·68	26·12	29·14	31·32
15	4·60	5·23	6·26	7·26	14·34	19·31	22·31	25·00	27·49	30·58	32·80
16	5·14	5·81	6·91	7·96	15·34	20·47	23·54	26·30	28·85	32·00	34·27
17	5·70	6·41	7·56	8·67	16·34	21·61	24·77	27·59	30·19	33·41	35·72
18	6·26	7·02	8·23	9·39	17·34	22·76	25·99	28·87	31·53	34·81	37·16
19	6·84	7·63	8·91	10·12	18·34	23·90	27·20	30·14	32·85	36·19	38·58
20	7·43	8·26	9·59	10·85	19·34	25·04	28·41	31·41	34·17	37·57	40·00
21	8·03	8·90	10·28	11·59	20·34	26·17	29·62	32·67	35·48	38·93	41·40
22	8·64	9·54	10·98	12·34	21·34	27·30	30·81	33·92	36·78	40·29	42·80
23	9·26	10·20	11·69	13·09	23·34	28·43	32·01	35·17	38·08	41·64	44·18
24	9·89	10·86	12·40	13·85	23·34	29·55	33·20	36·42	39·36	42·98	45·56
25	10·52	11·52	13·12	14·61	24·34	30·68	34·38	37·65	40·65	44·31	46·93
26	11·16	12·20	13·84	15·38	25·34	31·79	35·56	38·89	41·92	45·64	48·29
27	11·81	12·88	14·57	16·15	26·34	32·91	36·74	40·11	43·19	46·96	49·64
28	12·46	13·57	15·31	16·93	27·34	34·03	37·92	41·34	44·46	48·28	50·99
29	13·12	14·26	16·05	17·71	28·34	35·14	39·09	42·56	45·72	49·59	52·34
30	13·79	14·95	16·79	18·49	29·34	36·25	40·26	43·77	46·98	50·89	53·67
40	20·71	22·16	24·43	26·51	39·34	47·27	51·81	55·76	59·34	63·69	66·77
50	27·99	29·71	32·36	34·76	49·33	58·16	63·17	67·50	71·41	76·15	79·49
60	35·53	37·48	40·48	43·19	59·33	68·97	74·40	79·08	83·30	88·38	91·95
70	43·28	45·44	48·76	51·74	69·33	79·71	85·53	90·53	95·02	100·43	104·2
80	51·17	53·54	57·15	60·39	79·33	90·41	96·58	101·88	106·63	112·33	116·3
90	59·20	61·75	65·65	69·13	89·33	101·05	107·57	113·15	118·14	124·12	128·3
100	67·33	70·06	74·22	77·93	99·33	111·67	118·50	124·34	129·56	135·81	140·2

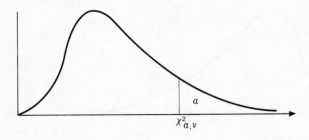

Table 4 Upper percentage points of the *F*-distribution

(a) $\alpha = 0.01$

ν_1	1	2	3	4	5	6	7	8	9	10	12	15	20	24	30
ν_2															
1	4052.2	4999.5	5403.4	5624.6	5763.6	5859.0	5928.4	5981.1	6022.5	6055.8	6106.3	6157.3	6208.7	6234.6	6260.6
2	98.50	99.00	99.17	99.25	99.30	99.33	99.36	99.37	99.39	99.40	99.42	99.43	99.45	99.46	99.47
3	34.12	30.82	29.46	28.71	28.24	27.91	27.67	27.49	27.35	27.23	27.05	26.87	26.69	26.60	26.50
4	21.20	18.00	16.69	15.98	15.52	15.21	14.98	14.80	14.66	14.55	14.37	14.20	14.02	13.93	13.84
5	16.26	13.27	12.06	11.39	10.97	10.67	10.46	10.29	10.16	10.05	9.89	9.72	9.55	9.47	9.38
6	13.75	10.92	9.78	9.15	8.75	8.47	8.26	8.10	7.98	7.87	7.72	7.56	7.40	7.31	7.23
7	12.25	9.55	8.45	7.85	7.46	7.19	6.99	6.84	6.72	6.62	6.47	6.31	6.16	6.07	5.99
8	11.26	8.65	7.59	7.01	6.63	6.37	6.18	6.03	5.91	5.81	5.67	5.52	5.36	5.28	5.20
9	10.56	8.02	6.99	6.42	6.06	5.80	5.61	5.47	5.35	5.26	5.11	4.96	4.81	4.73	4.65
10	10.04	7.56	6.55	5.99	5.64	5.39	5.20	5.06	4.94	4.85	4.71	4.56	4.41	4.33	4.25
11	9.65	7.21	6.22	5.67	5.32	5.07	4.89	4.74	4.63	4.54	4.40	4.25	4.10	4.02	3.94
12	9.33	6.93	5.95	5.41	5.06	4.82	4.64	4.50	4.39	4.30	4.16	4.01	3.86	3.78	3.70
13	9.07	6.70	5.74	5.21	4.86	4.62	4.44	4.30	4.19	4.10	3.96	3.82	3.66	3.59	3.51
14	8.86	6.51	5.56	5.04	4.69	4.46	4.28	4.14	4.03	3.94	3.80	3.66	3.51	3.43	3.35
15	8.68	6.36	5.42	4.89	4.56	4.32	4.14	4.00	3.89	3.80	3.67	3.52	3.37	3.29	3.21
16	8.53	6.23	5.29	4.77	4.44	4.20	4.03	3.89	3.78	3.69	3.55	3.41	3.26	3.18	3.10
17	8.40	6.11	5.18	4.67	4.34	4.10	3.93	3.79	3.68	3.59	3.46	3.31	3.16	3.08	3.00
18	8.29	6.01	5.09	4.58	4.25	4.01	3.84	3.71	3.60	3.51	3.37	3.23	3.08	3.00	2.92
19	8.18	5.93	5.01	4.50	4.17	3.94	3.77	3.63	3.52	3.43	3.30	3.15	3.00	2.92	2.84
20	8.10	5.85	4.94	4.43	4.10	3.87	3.70	3.56	3.46	3.37	3.23	3.09	2.94	2.86	2.78
21	8.02	5.78	4.87	4.37	4.04	3.81	3.64	3.51	3.40	3.31	3.17	3.03	2.88	2.80	2.72
22	7.95	5.72	4.82	4.31	3.99	3.76	3.59	3.45	3.35	3.26	3.12	2.98	2.83	2.75	2.67
23	7.88	5.66	4.76	4.26	3.94	3.71	3.54	3.41	3.30	3.21	3.07	2.93	2.78	2.70	2.62
24	7.82	5.61	4.72	4.22	3.90	3.67	3.50	3.36	3.26	3.17	3.03	2.89	2.74	2.66	2.58
25	7.77	5.57	4.68	4.18	3.85	3.63	3.46	3.32	3.22	3.13	2.99	2.85	2.70	2.62	2.54
26	7.72	5.53	4.64	4.14	3.82	3.59	3.42	3.29	3.18	3.09	2.96	2.81	2.66	2.58	2.50
27	7.68	5.49	4.60	4.11	3.78	3.56	3.39	3.26	3.15	3.06	2.93	2.78	2.63	2.55	2.47
28	7.64	5.45	4.57	4.07	3.75	3.53	3.36	3.23	3.12	3.03	2.90	2.75	2.60	2.52	2.44
29	7.60	5.42	4.54	4.04	3.73	3.50	3.33	3.20	3.09	3.00	2.87	2.73	2.57	2.49	2.41
30	7.56	5.39	4.51	4.02	3.70	3.47	3.30	3.17	3.07	2.98	2.84	2.70	2.55	2.47	2.39
40	7.31	5.18	4.31	3.83	3.51	3.29	3.12	2.99	2.89	2.80	2.66	2.52	2.37	2.29	2.20
60	7.08	4.98	4.13	3.65	3.34	3.12	2.95	2.82	2.72	2.63	2.50	2.35	2.20	2.12	2.03
120	6.85	4.79	3.95	3.48	3.17	2.96	2.79	2.66	2.56	2.47	2.34	2.19	2.03	1.95	1.86
∞	6.63	4.61	3.78	3.32	3.02	2.80	2.64	2.51	2.41	2.32	2.18	2.04	1.88	1.79	1.70

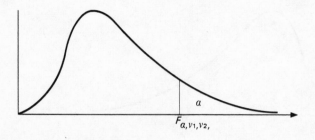

Table 4 (continued)

(b) $\alpha = 0.025$

v_1	1	2	3	4	5	6	7	8	9	10	12	15	20	24	30
v_2															
1	647·79	799·50	864·16	899·58	921·85	937·11	948·22	956·66	963·28	968·63	976·71	984·87	993·10	997·25	1001·4
2	38·51	39·00	39·17	39·25	39·30	39·33	39·36	39·37	39·39	39·40	39·41	39·43	39·45	39·46	39·46
3	17·44	16·04	15·44	1..10	14·88	14·73	14·62	14·54	14·47	14·42	14·34	14·25	14·17	14·12	14·08
4	12·22	10·65	9·98	9·60	9·36	9·20	9·07	8·98	8·90	8·84	8·75	8·66	8·56	8·51	8·46
5	10·01	8·43	7·76	7·39	7·15	6·98	6·85	6·76	6·68	6·62	6·52	6·43	6·33	6·28	6·23
6	8·81	7·26	6·60	6·23	5·99	5·82	5·70	5·60	5·52	5·46	5·37	5·27	5·17	5·12	5·07
7	8·07	6·54	5·89	5·52	5·29	5·12	4·99	4·90	4·82	4·76	4·67	4·57	4·47	4·41	4·36
8	7·57	6·06	5·42	5·05	4·82	4·65	4·53	4·43	4·36	4·30	4·20	4·10	4·00	3·95	3·89
9	7·21	5·71	5·08	4·72	4·48	4·32	4·20	4·10	4·03	3·96	3·87	3·77	3·67	3·61	3·56
10	6·94	5·46	4·83	4·47	4·24	4·07	3·95	3·85	3·78	3·72	3·62	3·52	3·42	3·37	3·31
11	6·72	5·26	4·63	4·28	4·04	3·88	3·76	3·66	3·59	3·53	3·43	3·33	3·23	3·17	3·12
12	6·55	5·10	4·47	4·12	3·89	3·73	3·61	3·51	3·44	3·37	3·28	3·18	3·07	3·02	2·96
13	6·41	4·97	4·35	4·00	3·77	3·60	3·48	3·39	3·31	3·25	3·15	3·05	2·95	2·89	2·84
14	6·30	4·86	4·24	3·89	3·66	3·50	3·38	3·29	3·21	3·15	3·05	2·95	2·84	2·79	2·73
15	6·20	4·77	4·15	3·80	3·58	3·41	3·29	3·20	3·12	3·06	2·96	2·86	2·76	2·70	2·64
16	6·12	4·69	4·08	3·73	3·50	3·34	3·22	3·12	3·05	2·99	2·89	2·79	2·68	2·63	2·57
17	6·04	4·62	4·01	3·66	3·44	3·28	3·16	3·06	2·98	2·92	2·82	2·72	2·62	2·56	2·50
18	5·98	4·56	3·95	3·61	3·38	3·22	3·10	3·01	2·93	2·87	2·77	2·67	2·56	2·50	2·44
19	5·92	4·51	3·90	3·56	3·33	3·17	3·05	2·96	2·88	2·82	2·72	2·62	2·51	2·45	2·39
20	5·87	4·46	3·86	3·51	3·29	3·13	3·01	2·91	2·84	2·77	2·68	2·57	2·46	2·41	2·35
21	5·83	4·42	3·82	3·48	3·25	3·09	2·97	2·87	2·80	2·73	2·64	2·53	2·42	2·37	2·31
22	5·79	4·38	3·78	3·44	3·22	3·05	2·93	2·84	2·76	2·70	2·60	2·50	2·39	2·33	2·27
23	5·75	4·35	3·75	3·41	3·18	3·02	2·90	2·81	2·73	2·67	2·57	2·47	2·36	2·30	2·24
24	5·72	4·32	3·72	3·38	3·15	2·99	2·87	2·78	2·70	2·64	2·54	2·44	2·33	2·27	2·21
25	5·69	4·29	3·69	3·35	3·13	2·97	2·85	2·75	2·68	2·61	2·51	2·41	2·30	2·24	2·18
26	5·66	4·27	3·67	3·33	3·10	2·94	2·82	2·73	2·65	2·59	2·49	2·39	2·28	2·22	2·16
27	5·63	4·24	3·65	3·31	3·08	2·92	2·80	2·71	2·63	2·57	2·47	2·36	2·25	2·19	2·13
28	5·61	4·22	3·63	3·29	3·06	2·90	2·78	2·69	2·61	2·55	2·45	2·34	2·23	2·17	2·11
29	5·59	4·20	3·61	3·27	3·04	2·88	2·76	2·67	2·59	2·53	2·43	2·32	2·21	2·15	2·09
30	5·57	4·18	3·59	3·25	3·03	2·87	2·75	2·65	2·57	2·51	2·41	2·31	2·20	2·14	2·07
40	5·42	4·05	3·46	3·13	2·90	2·74	2·62	2·53	2·45	2·39	2·29	2·18	2·07	2·01	1·94
60	5·29	3·93	3·34	3·01	2·79	2·63	2·51	2·41	2·33	2·27	2·17	2·06	1·94	1·88	1·82
120	5·15	3·80	3·23	2·89	2·67	2·52	2·39	2·30	2·22	2·16	2·05	1·94	1·82	1·76	1·69
∞	5·02	3·69	3·12	2·79	2·57	2·41	2·29	2·19	2·11	2·05	1·94	1·83	1·71	1·64	1·57

Table 4 (continued)

(c) $\alpha = 0.05$

ν_1	1	2	3	4	5	6	7	8	9	10	12	15	20	24	30
ν_2															
1	161·45	199·50	215·71	224·58	230·16	233·99	236·77	238·88	240·54	241·88	243·91	245·95	248·01	249·05	250·10
2	18·51	19·00	19·16	19·25	19·30	19·33	19·35	19·37	19·38	19·40	19·41	19·43	19·45	19·45	19·46
3	10·13	9·55	9·28	9·12	9·01	8·94	8·89	8·85	8·81	8·79	8·74	8·70	8·66	8·64	8·62
4	7·71	6·94	6·59	6·39	6·26	6·16	6·09	6·04	6·00	5·96	5·91	5·86	5·80	5·77	5·75
5	6·61	5·79	5·41	5·19	5·05	4·95	4·88	4·82	4·77	4·74	4·68	4·62	4·56	4·53	4·50
6	5·99	5·14	4·76	4·53	4·39	4·28	4·21	4·15	4·10	4·06	4·00	3·94	3·87	3·84	3·81
7	5·59	4·74	4·35	4·12	3·97	3·87	3·79	3·73	3·68	3·64	3·57	3·51	3·44	3·41	3·38
8	5·32	4·46	4·07	3·84	3·69	3·58	3·50	3·44	3·39	3·35	3·28	3·22	3·15	3·12	3·08
9	5·12	4·26	3·86	3·63	3·48	3·37	3·29	3·23	3·18	3·14	3·07	3·01	2·94	2·90	2·86
10	4·96	4·10	3·71	3·48	3·33	3·22	3·14	3·07	3·02	2·98	2·91	2·84	2·77	2·74	2·70
11	4·84	3·98	3·59	3·36	3·20	3·09	3·01	2·95	2·90	2·85	2·79	2·72	2·65	2·61	2·57
12	4·75	3·89	3·49	3·26	3·11	3·00	2·91	2·85	2·80	2·75	2·69	2·62	2·54	2·51	2·47
13	4·67	3·81	3·41	3·18	3·03	2·92	2·83	2·77	2·71	2·67	2·60	2·53	2·46	2·42	2·38
14	4·60	3·74	3·34	3·11	2·96	2·85	2·76	2·70	2·65	2·60	2·53	2·46	2·39	2·35	2·31
15	4·54	3·68	3·29	3·06	2·90	2·79	2·71	2·64	2·59	2·54	2·48	2·40	2·33	2·29	2·25
16	4·49	3·63	3·24	3·01	2·85	2·74	2·66	2·59	2·54	2·49	2·42	2·35	2·28	2·24	2·19
17	4·45	3·59	3·20	2·96	2·81	2·70	2·61	2·55	2·49	2·45	2·38	2·31	2·23	2·19	2·15
18	4·41	3·55	3·16	2·93	2·77	2·66	2·58	2·51	2·46	2·41	2·34	2·27	2·19	2·15	2·11
19	4·38	3·52	3·13	2·90	2·74	2·63	2·54	2·48	2·42	2·38	2·31	2·23	2·16	2·11	2·07
20	4·35	3·49	3·10	2·87	2·71	2·60	2·51	2·45	2·39	2·35	2·28	2·20	2·12	2·08	2·04
21	4·32	3·47	3·07	2·84	2·68	2·57	2·49	2·42	2·37	2·32	2·25	2·18	2·10	2·05	2·01
22	4·30	3·44	3·05	2·82	2·66	2·55	2·46	2·40	2·34	2·30	2·23	2·15	2·07	2·03	1·98
23	4·28	3·42	3·03	2·80	2·64	2·53	2·44	2·37	2·32	2·27	2·20	2·13	2·05	2·01	1·96
24	4·26	3·40	3·01	2·78	2·62	2·51	2·42	2·36	2·30	2·25	2·18	2·11	2·03	1·98	1·94
25	4·24	3·39	2·99	2·76	2·60	2·49	2·40	2·34	2·28	2·24	2·16	2·09	2·01	1·96	1·92
26	4·23	3·37	2·98	2·74	2·59	2·47	2·39	2·32	2·27	2·22	2·15	2·07	1·99	1·95	1·90
27	4·21	3·35	2·96	2·73	2·57	2·46	2·37	2·31	2·25	2·20	2·13	2·06	1·97	1·93	1·88
28	4·20	3·34	2·95	2·71	2·56	2·45	2·36	2·29	2·24	2·19	2·12	2·04	1·96	1·91	1·87
29	4·18	3·33	2·93	2·70	2·55	2·43	2·35	2·28	2·22	2·18	2·10	2·03	1·94	1·90	1·85
30	4·17	3·32	2·92	2·69	2·53	2·42	2·33	2·27	2·21	2·16	2·09	2·01	1·93	1·89	1·84
40	4·08	3·23	2·84	2·61	2·45	2·34	2·25	2·18	2·12	2·08	2·00	1·92	1·84	1·79	1·74
60	4·00	3·15	2·76	2·53	2·37	2·25	2·17	2·10	2·04	1·99	1·92	1·84	1·75	1·70	1·65
120	3·92	3·07	2·68	2·45	2·29	2·18	2·09	2·02	1·96	1·91	1·83	1·75	1·66	1·61	1·55
∞	3·84	3·00	2·60	2·37	2·21	2·10	2·01	1·94	1·88	1·83	1·75	1·67	1·57	1·52	1·46

Table 5 Values of e^{-x}

x	$e^{-0.01x}$	$e^{-0.1x}$	e^{-x}	x	$e^{-0.01x}$	$e^{-0.1x}$	e^{-x}
0·0	1·0000	1·0000	1·00000				
·1	·9990	·9900	·90484	5·1	·9503	·6005	·00610
·2	·9980	·9802	·81873	5·2	·9493	·5945	·00552
·3	·9970	·9704	·74082	5·3	·9484	·5886	·00499
·4	·9960	·9608	·67032	5·4	·9474	·5827	·00452
·5	·9950	·9512	·60653	5·5	·9465	·5769	·00409
·6	·9940	·9418	·54881	5·6	·9455	·5712	·00370
·7	·9930	·9324	·49659	5·7	·9446	·5655	·00335
·8	·9920	·9231	·44933	5·8	·9436	·5599	·00303
·9	·9910	·9139	·40657	5·9	·9427	·5543	·00274
1·0	·9900	·9048	·36788	6·0	·9418	·5488	·00248
1·1	·9891	·8958	·33287	6·1	·9408	·5434	·00224
1·2	·9881	·8869	·30119	6·2	·9399	·5379	·00203
1·3	·9871	·8781	·27253	6·3	·9389	·5326	·00184
1·4	·9861	·8694	·24660	6·4	·9380	·5273	·00166
1·5	·9851	·8607	·22313	6·5	9371	·5220	·00150
1·6	·9841	·8521	·20190	6·6	·9361	·5169	·00136
1·7	·9831	·8437	·18268	6·7	·9352	·5117	·00123
1·8	·9822	·8353	·16530	6·8	·9343	·5066	·00111
1·9	·9812	·8270	·14957	6·9	·9333	·5016	·00101
2·0	·9802	·8187	·13534	7·0	·9324	·4966	·00091
2·1	·9792	·8106	·12246	7·1	·9315	·4916	·00083
2·2	·9782	·8025	·11080	7·2	·9305	·4868	·00075
2·3	·9773	·7945	·10026	7·3	·9296	·4819	·00068
2·4	·9763	·7866	·09072	7·4	·9287	·4771	·00061
2·5	·9753	·7788	·08208	7·5	·9277	·4724	·00055
2·6	·9743	·7711	·07427	7·6	·9268	·4677	·00050
2·7	·9734	·7634	·06721	7·7	·9259	·4630	·00045
2·8	·9724	·7558	·06081	7·8	·9250	·4584	·00041
2·9	·9714	·7483	·05502	7·9	·9240	·4538	·00037
3·0	·9704	·7408	·04979	8·0	·9231	·4493	·00034
3·1	·9695	·7334	·04505	8·1	·9222	·4449	·00030
3·2	·9685	·7261	·04076	8·2	·9213	·4404	·00027
3·3	·9675	·7189	·03688	8·3	·9204	·4360	·00025
3·4	·9666	·7118	·03337	8·4	·9194	·4317	·00022
3·5	·9656	·7047	·03020	8·5	·9185	·4274	·00020
3·6	·9646	·6977	·02732	8·6	·9176	·4232	·00018
3·7	·9637	·6907	·02472	8·7	·9167	·4190	·00017
3·8	·9627	·6839	·02237	8·8	·9158	·4148	·00015
3·9	·9618	·6771	·02024	8·9	·9148	·4107	·00014
4·0	·9608	·6703	·01832	9·0	·9139	·4066	·00012
4·1	·9598	·6637	·01657	9·1	·9130	·4025	·00011
4·2	·9589	·6570	·01500	9·2	·9121	·3985	·00010
4·3	·9579	·6505	·01357	9·3	·9112	·3946	·00009
4·4	·9570	·6440	·01228	9·4	·9103	·3906	·00008
4·5	·9560	·6376	·01111	9·5	·9094	·3867	·00007
4·6	·9550	·6313	·01005	9·6	·9085	·3829	·00007
4·7	·9541	·6250	·00910	9·7	·9076	·3791	·00006
4·8	·9531	·6188	·00823	9·8	·9066	·3753	·00006
4·9	·9522	·6126	·00745	9·9	·9057	·3716	·00005
5·0	·9512	·6065	·00674	10·0	·9048	·3679	·00005

Table 6 Percentage points of the distribution of the Studentized range

(a) $\alpha = 0.01$

c	2	3	4	5	6	7	8	9	10	11	12	13	14	15	16	17	18	19	20
ν																			
1	90·0	135	164	186	202	216	227	237	246	253	260	266	272	277	282	286	290	294	298
2	14·0	19·0	22·3	24·7	26·6	28·2	29·5	30·7	31·7	32·6	33·4	34·1	34·8	35·4	36·0	36·5	37·0	37·5	37·9
3	8·26	10·6	12·2	13·3	14·2	15·0	15·6	16·2	16·7	17·1	17·5	17·9	18·2	18·5	18·8	19·1	19·3	19·5	19·8
4	6·51	8·12	9·17	9·96	10·6	11·1	11·5	11·9	12·3	12·6	12·8	13·1	13·3	13·5	13·7	13·9	14·1	14·2	14·4
5	5·70	6·97	7·80	8·42	8·91	9·32	9·67	9·97	10·24	10·48	10·70	10·89	11·08	11·24	11·40	11·55	11·68	11·81	11·93
6	5·24	6·33	7·03	7·56	7·97	8·32	8·61	8·87	9·10	9·30	9·49	9·65	9·81	9·95	10·08	10·21	10·32	10·43	10·54
7	4·95	5·92	6·54	7·01	7·37	7·68	7·94	8·17	8·37	8·55	8·71	8·86	9·00	9·12	9·24	9·35	9·46	9·55	9·65
8	4·74	5·63	6·20	6·63	6·96	7·24	7·47	7·68	7·87	8·03	8·18	8·31	8·44	8·55	8·66	8·76	8·85	8·94	9·03
9	4·60	5·43	5·96	6·35	6·66	6·91	7·13	7·32	7·49	7·65	7·78	7·91	8·03	8·13	8·23	8·32	8·41	8·49	8·57
10	4·48	5·27	5·77	6·14	6·43	6·67	6·87	7·05	7·21	7·36	7·48	7·60	7·71	7·81	7·91	7·99	8·07	8·15	8·22
11	4·39	5·14	5·62	5·97	6·25	6·48	6·67	6·84	6·99	7·13	7·25	7·36	7·46	7·56	7·65	7·73	7·81	7·88	7·95
12	4·32	5·04	5·50	5·84	6·10	6·32	6·51	6·67	6·81	6·94	7·06	7·17	7·26	7·36	7·44	7·52	7·59	7·66	7·73
13	4·26	4·96	5·40	5·73	5·98	6·19	6·37	6·53	6·67	6·79	6·90	7·01	7·10	7·19	7·27	7·34	7·42	7·48	7·55
14	4·21	4·89	5·32	5·63	5·88	6·08	6·26	6·41	6·54	6·66	6·77	6·87	6·96	7·05	7·12	7·20	7·27	7·33	7·39
15	4·17	4·83	5·25	5·56	5·80	5·99	6·16	6·31	6·44	6·55	6·66	6·76	6·84	6·93	7·00	7·07	7·14	7·20	7·26
16	4·13	4·78	5·19	5·49	5·72	5·92	6·08	6·22	6·35	6·46	6·56	6·66	6·74	6·82	6·90	6·97	7·03	7·09	7·15
17	4·10	4·74	5·14	5·43	5·66	5·85	6·01	6·15	6·27	6·38	6·48	6·57	6·66	6·73	6·80	6·87	6·94	7·00	7·05
18	4·07	4·70	5·09	5·38	5·60	5·79	5·94	6·08	6·20	6·31	6·41	6·50	6·58	6·65	6·72	6·79	6·85	6·91	6·96
19	4·05	4·67	5·05	5·33	5·55	5·73	5·89	6·02	6·14	6·25	6·34	6·43	6·51	6·58	6·65	6·72	6·78	6·84	6·89
20	4·02	4·64	5·02	5·29	5·51	5·69	5·84	5·97	6·09	6·19	6·29	6·37	6·45	6·52	6·59	6·65	6·71	6·76	6·82
24	3·96	4·54	4·91	5·17	5·37	5·54	5·69	5·81	5·92	6·02	6·11	6·19	6·26	6·33	6·39	6·45	6·51	6·56	6·61
30	3·89	4·45	4·80	5·05	5·24	5·40	5·54	5·65	5·76	5·85	5·93	6·01	6·08	6·14	6·20	6·26	6·31	6·36	6·41
40	3·82	4·37	4·70	4·93	5·11	5·27	5·39	5·50	5·60	5·69	5·77	5·84	5·90	5·96	6·02	6·07	6·12	6·17	6·21
60	3·76	4·28	4·60	4·82	4·99	5·13	5·25	5·36	5·45	5·53	5·60	5·67	5·73	5·79	5·84	5·89	5·93	5·98	6·02
120	3·70	4·20	4·50	4·71	4·87	5·01	5·12	5·21	5·30	5·38	5·44	5·51	5·56	5·61	5·66	5·71	5·75	5·79	5·83
∞	3·64	4·12	4·40	4·60	4·76	4·88	4·99	5·08	5·16	5·23	5·29	5·35	5·40	5·45	5·49	5·54	5·57	5·61	5·65

c is the size of the sample from which the range is obtained and ν is the number of degrees of freedom of s.

(b) $a = 0.05$

c	2	3	4	5	6	7	8	9	10	11	12	13	14	15	16	17	18	19	20
v																			
1	18·0	27·0	32·8	37·1	40·4	43·1	45·4	47·4	49·1	50·6	52·0	53·2	54·3	55·4	56·3	57·2	58·0	58·8	59·6
2	6·09	8·3	9·8	10·9	11·7	12·4	13·0	13·5	14·0	14·4	14·7	15·1	15·4	15·7	15·9	16·1	16·4	16·6	16·8
3	4·50	5·91	6·82	7·50	8·04	8·48	8·85	9·18	9·46	9·72	9·95	10·15	10·35	10·52	10·69	10·84	10·98	11·11	11·24
4	3·93	5·04	5·76	6·29	6·71	7·05	7·35	7·60	7·83	8·03	8·21	8·37	8·52	8·66	8·79	8·91	9·03	9·13	9·23
5	3·64	4·60	5·22	5·67	6·03	6·33	6·58	6·80	6·90	7·17	7·32	7·47	7·60	7·72	7·83	7·93	8·03	8·12	8·21
6	3·46	4·34	4·90	5·31	5·63	5·89	6·12	6·32	6·49	6·65	6·79	6·92	7·03	7·14	7·24	7·34	7·43	7·51	7·59
7	3·34	4·16	4·68	5·06	5·36	5·61	5·82	6·00	6·16	6·30	6·43	6·55	6·66	6·76	6·85	6·94	7·02	7·09	7·17
8	3·26	4·04	4·53	4·89	5·17	5·40	5·60	5·77	5·92	6·05	6·18	6·29	6·39	6·48	6·57	6·65	6·73	6·80	6·87
9	3·20	3·95	4·42	4·76	5·02	5·24	5·43	5·60	5·74	5·87	5·98	6·09	6·19	6·28	6·36	6·44	6·51	6·58	6·64
10	3·15	3·88	4·33	4·65	4·91	5·12	5·30	5·46	5·60	5·72	5·83	5·93	6·03	6·11	6·20	6·27	6·34	6·40	6·47
11	3·11	3·82	4·26	4·57	4·82	5·03	5·20	5·35	5·49	5·61	5·71	5·81	5·90	5·99	6·06	6·14	6·20	6·26	6·33
12	3·08	3·77	4·20	4·51	4·75	4·95	5·12	5·27	5·40	5·51	5·62	5·71	5·80	5·88	5·95	6·03	6·09	6·15	6·21
13	3·06	3·73	4·15	4·45	4·69	4·88	5·05	5·19	5·32	5·43	5·53	5·63	5·71	5·79	5·86	5·93	6·00	6·05	6·11
14	3·03	3·70	4·11	4·41	4·64	4·83	4·99	5·13	5·25	5·36	5·46	5·55	5·64	5·72	5·79	5·85	5·92	5·97	6·03
15	3·01	3·67	4·08	4·37	4·60	4·78	4·94	5·08	5·20	5·31	5·40	5·49	5·58	5·65	5·72	5·79	5·85	5·90	5·96
16	3·00	3·65	4·05	4·33	4·56	4·74	4·90	5·03	5·15	5·26	5·35	5·44	5·52	5·59	5·66	5·72	5·79	5·84	5·90
17	2·98	3·63	4·02	4·30	4·52	4·71	4·86	4·99	5·11	5·21	5·31	5·39	5·47	5·55	5·61	5·68	5·74	5·79	5·84
18	2·97	3·61	4·00	4·28	4·49	4·67	4·82	4·96	5·07	5·17	5·27	5·35	5·43	5·50	5·57	5·63	5·69	5·74	5·79
19	2·96	3·59	3·98	4·25	4·47	4·65	4·79	4·92	5·04	5·14	5·23	5·32	5·39	5·46	5·53	5·59	5·65	5·70	5·75
20	2·95	3·58	3·96	4·23	4·45	4·62	4·77	4·90	5·01	5·11	5·20	5·28	5·36	5·43	5·49	5·55	5·61	5·66	5·71
24	2·92	3·53	3·90	4·17	4·37	4·54	4·68	4·81	4·92	5·01	5·10	5·18	5·25	5·32	5·38	5·44	5·50	5·54	5·59
30	2·89	3·49	3·84	4·10	4·30	4·46	4·60	4·72	4·83	4·92	5·00	5·08	5·15	5·21	5·27	5·33	5·38	5·43	5·48
40	2·86	3·44	3·79	4·04	4·23	4·39	4·52	4·63	4·74	4·82	4·91	4·98	5·05	5·11	5·16	5·22	5·27	5·31	5·36
60	2·83	3·40	3·74	3·98	4·16	4·31	4·44	4·55	4·65	4·73	4·81	4·88	4·94	5·00	5·06	5·11	5·16	5·20	5·24
120	2·80	3·36	3·69	3·92	4·10	4·24	4·36	4·48	4·56	4·64	4·72	4·78	4·84	4·90	4·95	5·00	5·05	5·09	5·13
∞	2·77	3·31	3·63	3·86	4·03	4·17	4·29	4·39	4·47	4·55	4·62	4·68	4·74	4·80	4·85	4·89	4·93	4·97	5·01

c is the size of the sample from which the range is obtained and v is the number of degrees of freedom of s.

Table 7 Random numbers

19211	73336	80586	08681	28012	48881	34321	40156	03776	45150
94520	44451	07032	36561	41311	28421	95908	91280	74627	86359
70986	03817	40251	61310	25940	92411	34796	85416	00993	99487
65249	79677	03155	09232	96784	17126	50350	86469	41300	62715
82102	03098	01785	00653	39438	43660	02406	08404	24540	80000
91600	94635	35392	81737	01505	04967	91097	02011	26642	38540
20559	85361	20093	46000	83304	96624	62541	41722	79676	98970
53305	79544	99937	87727	32210	19438	58250	77265	02998	02973
57108	86498	14158	60697	41673	18087	46088	11238	82135	79035
08270	11929	92040	37390	71190	58952	98702	41638	95725	22798
90119	23206	75634	60053	90724	29080	69423	66815	11896	18607
45124	69607	17078	61747	15891	69904	79589	68137	19006	19045
83084	02589	37660	63882	99025	34831	92048	23671	68895	73795
04685	31035	93828	16159	05015	54800	76534	22974	13589	01801
61349	04538	89318	27693	02674	34368	24720	40682	20940	37392
14082	65020	49956	01336	41685	01758	49242	52122	01030	60378
82615	53477	58014	62229	72640	32042	73521	14166	45850	02372
50942	78633	16588	19275	62258	20773	67601	93065	69002	03985
76381	77455	81218	02520	22900	80130	61554	98901	26939	78732
05645	35063	85932	22410	31357	54790	39707	94348	11969	89755
76591	83750	46137	74989	39931	33068	35155	49486	28156	04556
31945	87960	04852	41411	63105	44116	95250	04046	59211	67270
08648	89822	04170	38365	23842	61917	57453	03495	61430	20154
32511	07999	18920	77045	44299	85057	51395	17457	24207	02730
79348	56194	58145	88645	84867	41594	28148	84985	89949	26689
61973	03660	32988	70689	17794	61340	58311	32569	23949	85626
92032	60127	34066	28149	22352	12907	53788	86648	57649	07887
74609	71072	63958	58336	67814	40598	12626	30754	75895	42194
98668	76074	25634	56913	88254	41647	05398	69463	49778	31382
65248	72078	58634	88678	21764	67940	45666	84664	35714	43081
82002	96916	94138	74739	99122	03904	46052	97277	60243	37424
79100	55938	23211	10111	17115	90577	94202	01063	85522	64378
30923	71710	70257	05596	42310	02449	31211	50025	99744	78084
90513	50966	78981	70391	45932	13535	21681	66589	94915	08855
94474	79356	16098	95806	79252	14190	88722	39887	15553	58386
65236	62948	19968	22071	49898	96140	80264	57580	56775	63138
80502	04192	84287	32589	50664	63846	71590	67220	71503	27942
01315	04632	50202	89148	41556	11584	35916	13979	25016	32511
81525	76670	88714	28681	56540	84963	85543	69715	86192	79373
19500	41720	79214	20079	42053	29844	02294	11306	78537	65098
25812	77090	45198	98162	13782	60596	99092	50188	65405	63227
80859	94220	92309	01998	45090	24815	13415	86989	01677	39002
41107	33561	04376	40072	78909	61042	04098	73304	21892	63112
00465	00858	22774	80730	07098	80515	09970	40476	10314	24792
58137	02454	15657	24957	48401	02940	92828	26372	31071	58192
32013	97147	69725	78867	73329	74935	69276	46001	04181	38838
17048	84788	12531	01773	43551	34586	61239	87927	03232	31312
33935	07944	98456	11922	96174	24100	00307	85697	06527	34381
47633	49394	38673	22281	68096	76599	38462	16662	81959	03358
82161	92521	10712	58839	18546	32920	89220	90493	73725	22327

Table 7 (continued)

99050	30876	80821	14955	11495	25666	37656	91874	93051	64664
08090	84688	36332	86858	73763	62534	93378	54809	97076	09077
67619	00352	32735	56954	97851	57350	33068	35393	75938	86086
63779	66008	02516	93878	67930	38445	44166	20168	55128	65337
03259	72119	04797	95593	02754	87120	68167	04455	75318	93127
92914	02066	97320	00328	51685	89729	27446	32599	82486	01718
80001	70542	01530	63033	64348	01306	75419	90348	34717	05147
38715	09824	86504	14817	74434	80450	95086	73824	40550	14266
15987	74578	12779	69608	76893	94840	36853	00568	35697	00783
06193	94893	24598	02714	69670	06153	97835	71087	58193	97912
40134	12803	33942	46660	05681	35209	65980	77899	38988	75580
88480	27598	48458	65369	81066	02000	68719	90488	50062	10428
49989	94369	80429	97152	67032	62342	96496	91274	71264	45271
62089	52111	92190	85413	95362	33400	03488	84666	99974	01459
01675	12741	94334	86069	71353	85566	16632	97577	18708	99550
04529	19798	47711	63262	06316	00287	86718	33705	31645	70615
63895	63087	91886	43467	55559	35912	39429	18933	75931	18924
17709	21642	56384	85699	24310	85043	00405	59820	54228	58645
11727	83872	22553	17012	02849	39794	50662	32647	67676	95488
02838	03160	92864	29385	63585	46055	41356	96398	70904	87103
62210	02385	73776	03171	83842	94602	31540	96071	55024	87629
16825	05535	99451	81864	99410	81211	62781	55121	62268	48522
05985	62766	58215	61900	53065	85082	88200	74393	24100	88379
14184	86400	41788	82932	27183	44744	14964	71718	76499	37364
95315	04537	85490	90542	42519	35659	87983	51941	20420	56828
65578	64820	95644	98074	72032	53443	92722	96373	36030	78053
18444	28477	01846	95805	91166	74383	55926	92971	99743	04905
03577	99361	21047	21971	71191	70493	70210	87051	94715	88924
49752	47015	09472	20089	90924	03674	73181	81104	95411	00656
32489	04936	30628	99512	40891	39832	25101	71757	77503	82112
76548	92824	53738	65890	78297	50705	96792	56841	41063	92875
26545	68726	06476	57444	35455	46706	40388	79728	99747	75076
67651	97346	75509	50270	27943	71144	15397	04565	95265	52236
67879	04880	01478	97239	32611	85024	37275	46399	59303	28341
96329	85824	79954	96263	91873	37394	45728	32769	72930	82361
87421	32587	32890	79171	54734	60628	53702	06741	98558	19167
22447	88823	21866	39773	26018	28765	01876	03776	51523	89095
79589	92914	06964	43330	01726	30504	24797	52657	44098	22006
92123	79976	31751	68549	06147	38138	58792	80966	59767	24564
85909	35590	89231	75271	34409	48770	08980	54457	26022	29742
43162	44793	39006	76661	02000	14571	73986	96351	02276	47746
47549	41709	52412	40595	40397	38883	20843	90121	74897	96286
71711	75690	50441	41322	16497	36962	88880	45374	29836	82096
51091	24078	13706	27315	69918	06628	99964	09477	59496	90825
94981	73799	35590	58944	36581	94509	17508	31203	97030	28541
23778	02351	44843	28005	63835	69611	91360	20756	70188	02554
36324	01285	47959	40386	10284	03089	95441	77955	70381	04689
31710	55804	18079	15172	27321	93535	81303	97488	94531	61924
84106	55010	57902	09150	59719	52718	96632	22555	72411	85957
27527	60618	02688	95261	20022	88691	20488	93189	33658	49237

Further Reading

A list of general reference books will be given here, together with several books on important topics which have not been covered in this book.

General reference books

The following list gives a selection of general reference books on different aspects of statistics. The books edited by Davies are especially relevant to problems in the chemical industry. A mathematical approach to statistics is provided by Hoel. An elementary introduction to statistical methods is provided by Wetherill. Snedecor and Cochran give a comprehensive coverage of statistical methods. The two most commonly used sets of statistical tables are Fisher and Yates, and Pearson and Hartley.

R. L. ANDERSON and T. A. BANCROFT, *Statistical Theory in Research*, McGraw-Hill, 1952.

C. A. BENNETT and N. L. FRANKLIN, *Statistical Analysis in Chemistry and the Chemical Industry*, Wiley, 1954.

A. H. BOWKER and G. J. LIEBERMANN, *Engineering Statistics*, Prentice-Hall, 1957.

K. A. BROWNLEE, *Statistical Theory and Methodology in Science and Engineering*, Wiley, 1960.

O. L. DAVIES (ed.), *Design and Analysis of Industrial Experiments*, Oliver & Boyd, 3rd edn, 1957.

O. L. DAVIES (ed.), *Statistical Methods in Research and Production*, Oliver & Boyd, 2nd edn, 1960.

R. A. FISHER and F. YATES, *Statistical Tables of Biological, Agricultural and Medical Research*, Oliver & Boyd, 6th edn, 1963.

A. HALD, *Statistical Theory with Engineering Applications*, Wiley, 1952.

P. G. HOEL, *Introduction to Mathematical Statistics*, Wiley, 3rd edn, 1964.

M. G. KENDALL and A. STUART, *The Advanced Theory of Statistics*, vol. 1, Griffin, 2nd edn, 1963.

E. S. PAGE, 'Cumulative sum charts' *Technometrics*, vol. 3 (1961), pp. 1–10.

E. S. Pearson and H. O. Hartley, *Biometrika Tables for Statisticians*, Cambridge University Press, 3rd edn, 1966.

G. W. Snedecor and W. G. Cochran, *Statistical Methods*, Iowa State University Press, 6th edn, 1967.

G. B. Wetherill, *Elementary Statistical Methods*, Methuen, 1967.

R. L. Wine, *Statistics for Scientists and Engineers*, Prentice-Hall, 1964.

Operations research

There is a group of topics, of particular interest to the production engineer, which is often given the collective title operations research. Some of these topics are related to probability and statistics but, for reasons of space, no attempt has been made to describe them in this book. The topics include simulation, inventory control, linear programming, dynamic programming, game theory, queueing theory and critical path scheduling.

Simulation is an important tool for solving many types of production problem. The technique involves the construction of a mathematical model of the relevant portion of the production system and this enables experiments to be made with the system on paper (or more usually on a computer). This avoids costly experimental changes to the actual system. Monte Carlo simulation involves experiments based on random numbers.

Linear programming is a technique for allocating some limited resources to competing demands in the most effective way.

W. D. Ashton, *The Theory of Road Traffic Flow*, Methuen, 1966.

G. A. Barnard, 'Control charts and stochastic processes', *Journal of the Royal Statistical Society*, series B. vol. 21 (1959), no. 2, pp. 239–71.

Z. W. Birnbaum and S. C. Saunders, 'A statistical model for the life-length of materials', *Journal of the American Statistical Association*, vol. 53 (1958), pp. 151–60.

E. S. Buffa, *Models for Production and Operations Management*, Wiley, 1963.

C. W. Churchman, R. L. Ackoff and E. L. Arnoff, *Introduction to Operations Research*, Wiley, 1957.

D. R. Cox and W. L. Smith, *Queues*, Wiley, 1961.

S. Eilon, *Elements of Production Planning and Control*, Macmillan, 1962.

W. D. Ewan, 'When and how to use cusum charts', *Technometrics*, vol. 5 (1963), pp. 1–22.

Z. Govindarajula, 'Supplement to Mendenhall's bibliography', *Journal of the American Statistical Association*, vol. 59 (1964), pp. 1231–91.

J. M. Hammersley and D. C. Handscomb, *Monte Carlo Methods*, Methuen, 1964.

J. HOROWITZ, *Critical Path Scheduling*, Ronald Press Co., 1967.

W. MENDENHALL, 'A bibliography on life-testing and related topics', *Biometrika*, vol. 45 (1958), pp. 521–43.

T. H. NAYLOR, J. L. BALINTFY, D. S. BURDICK and K. CHU, *Computer Simulation Techniques*, Wiley, 1966.

E. S. PAGE, 'Cumulative sum charts', *Technometrics*, vol. 3 (1961), pp. 1–10.

M. SASIENI, A. YASPAN and L. FRIEDMAN, *Operations Research*, Wiley, 1959.

L. TAKACS, *Introduction to the Theory of Queues*, Oxford University Press, 1962.

K. D. TOCHER, *The Art of Simulation*, English University Press, 1963.

Some other topics

There are a number of topics of interest to engineers which are related to probability rather than to statistics. These require a more thorough treatment of probability than that attempted here. They also involve a knowledge of stochastic processes, which are physical processes whose structure involves a random mechanism. An example of such a process is the Poisson process described in Chapter 4. The topics include control theory, the statistical theory of communication including information theory, and time-series analysis.

R. B. BLACKMAN and J. W. TUKEY, *The Measurement of Power Spectra*, Dover, 1959.

W. FELLER, *An Introduction to Probability Theory and its Applications, vol. 1*, Wiley, 3rd edn, 1968.

G. M. JENKINS and D. G. WATTS, *Spectral Analysis and its Application*, Holden-Day, 1968.

D. MIDDLETON, *An Introduction to Statistical Communication Theory*, McGraw-Hill, 1960.

A. PAPOULIS, *Probability, Random Variables and Stochastic Processes*, McGraw-Hill, 1965.

E. PARZEN, *Modern Probability Theory and its Applications*, Wiley, 1960.

Answers to exercises

Chapter 2

2. (a) 4·6, 5·3, 2·3 (b) 7·0, 54·5, 7·3
3. (a) 995·4, 5·3 (b) 130 is a suitable constant, $\bar{x} = 136·2$, $s^2 = 11·8$

Chapter 3

1. (a) $(1 - p_1)(1 - p_2)$ (b) $1 - (1 - p_1)(1 - p_2)$
(c) $1 - (1 - p_1)(1 - p_2) - p_1 p_2$
2. 0·99
3. (a) $\frac{1}{6}$ (b) $\frac{15}{36}$ (c) $\frac{1}{6}$ (look at sample space)
4. (a) $\frac{5}{16}$ (b) $\frac{11}{16}$ (write out all possible outcomes)
5. (a) $\frac{3}{5}$ (b) $\frac{3}{5}$ (c) $\frac{3}{10}$
6. 60×59
7. 26^3
8. 23
9. 0·59
10. (a) 0·72 (b) 0·921 (c) 0·950
11. $1 - (\frac{5}{6})^6 = 0·67$

Chapter 4

1. Binomial distribution; (a) 0·2 (b) 0·817 (c) 0·017
2. (b) Expand $(1 + 1)^n$ (c) Expand $(1 - 1)^n$
3. Number of defectives follows a binomial distribution with $n = 10$ and $p = 0·4$; 5·6%; very grave doubts – take another sample immediately.
4. $n \geqslant 70$
5. 0·0085
6. Yes, P(2 additional failures) $\approx 0·00000003$
Yes, P(2 additional failures) $\approx 0·00000001$
So three-engine jet is three times as reliable in this situation.

7. (a) 10 cars (b) Yes

8. Two; third ambulance would be used once in 147 hours; a good idea.

9. 0·9989; ignore second signal and pray.

10. (a) 0·367 (b) 0·368

11. (a) 0·135 (b) 0·325

Chapter 5

1. (a) 0·317 (b) 0·383 (c) 0·036

2. (a) 0·5 (b) 0·841 (c) 0·159

3. (a) 0·954 (b) 0·023

4. 0·017

5. (a) 13·3 (b) 14·7; $k = 3·92$

6. 97·7% will have a life-time > 400 hours. Therefore purchaser will be satisfied.

7. 12 questions

8. Expected proportion of bolts less than 2·983 in. $= 0·015; E(n_1) = 4·5$, $E(n_2) = 30·6$. In second case solve $\mu - 1·84\sigma = 2·983$ and $\mu + 1·96\sigma = 3·021$, giving $\mu = 3·001, \sigma = 0·010$.

Chapter 6

1. (a) 0·66 (b) 0·4 (c) 0·2; sample size $= 16$

3. $a = -0·52, b = 0·52, c = 0·37, d = 1·88$

$a = 4·95, b = 7·05, c = 1·48, d = 7·52$

5. $14·1 \pm 2·0$

6. $14·1 \pm 2·2$

Chapter 7

1. $z_0 = 1·33$; two-tailed level of significance $> 5\%$; therefore accept H_0. 95% confidence interval for μ is $12·22 \pm 1·96 \times 0·4/\sqrt{6}$; note that this interval contains 12·0.

2. $s = 0·37$, $t_0 = 1·45$; two-tailed level of significance $> 5\%$; accept H_0. 95% confidence interval for μ is $12·22 \pm 2·57 \times 0·37/\sqrt{6}$.

3. $H_0 : \mu = 3\%$; $H_1 : \mu < 3\%$; one-tailed test; $t_0 = 3·16$. But $t_{0·01,9} = 2·82$; therefore reject H_0 at 1% level.

4. n is smallest integer above $3^2(1·64 + 1·64)^2/(32 - 30)^2$; $n = 25$

5. $t_0 = 4.74$; significant at 1% level
6. $s^2 = 34.5$, $t_0 = 2.49$; (a) Yes (b) No
7. $\chi_0^2 = 7.0$; 9 degrees of freedom; accept hypothesis
8. Combine three or more accidents; $\chi_0^2 = 61.1$ with 2 d.f.; reject hypothesis. (Some women are accident prone so that the Poisson parameter is not the same for all members of the population. See M. G. Kendall and A. Stuart, *Advanced theory of Statistics, Vol. 1*, Griffin, 2nd edn, 1963.)
9. $\chi_0^2 = 2.82$ with 1 d.f.; not significantly large; effectiveness of drug has not been proved.

Chapter 8

1. $\hat{y} = 0.514 + 0.0028x$; $\hat{y} = 0.584$
2. $\hat{y} = 13.46 + 1.13 \times \text{coded year}$; $\hat{y} = 16.85$
3. (a) 0.780 (b) weight $= 81.2 + 1.04 \times \text{height}$
(c) height $= -19.1 + 0.579 \times \text{weight}$
4. $\hat{y} = -0.073 + 0.940x_1 + 0.597x_2$

Chapter 9

2. $\sigma_z^2 \approx \left(\dfrac{1}{3x_3}\right)^2 \sigma_{x_1}^2 + \left(\dfrac{1}{3x_3}\right)^2 \sigma_{x_2}^2 + \left(\dfrac{x_1 + x_2}{3x_3^2}\right)^2 \sigma_{x_3}^2 = 0.381\sigma_{x_3}^2$

3. $\dfrac{(m_1 + m_2 + m_3 + 3m_4)}{4}$

4. $\dfrac{25m_1}{34} + \dfrac{9m_2}{34}$

5. $E(z) \approx 1.06 \times 2.31$; $\sigma_z^2 \approx (0.04)^2 \times (2.31)^2 + (1.06)^2 \times (0.05)^2$; tolerance limits are 2.45 ± 0.32

Chapter 10

4. $F = 5.4$; significant at 5% level
5. $F = 11.4$; significant at 1% level
6. $F = 77$; significant at 1% level
7. Ten blocks are necessary: ABC, ABD, ABE, ACD, ACE, ADE, BCD, BCE, BDE, CDE; randomize the order within each block.

Chapter 11

2. ANOVA *table*

Source	Sum of squares	d.f.	Mean square	F
Catalysts	117·8	3	39·27	20·1
Agitation	90·2	2	45·1	23·1
Interaction	4·1	6	0·68	0·4
Blocks	0·04	1	0·04	0·02
Residual	21·5	11	1·95	
Total	233·6	23		

Catalyst and agitation effects are highly significant; interaction not significant. By inspection CE_3 is best combination, but this is only marginally better than CE_2 which would probably be preferred on economic grounds.

3. $A: -3\frac{1}{2}$, $B: 0$, $C: -7$, $AB: -2$, $AC: -1$, $BC: \frac{1}{2}$, $ABC: -\frac{1}{2}$; A effect significant at 5% level, B effect not significant, C effect significant at 1% level

Chapter 12

1. $\bar{x}$-chart has action lines at $50 \pm 15/(2·32 \times \sqrt{5})$; R-chart has action limit at $5 \times 5·48/2·32$.

2. $\bar{x}$-chart has action lines at $10·80 \pm 3 \times 0·46/(2·06 \times 2)$; R-chart has action limit at $0·46 \times 5·31/2·06$.

3. $P(r$ defectives in sample size 30$) \approx e^{-0·9}(0·9)^r/r!$; warning limit $= 3$ defectives; action limit $= 5$ defectives

4. $0·023$;
$P(r$ observations taken before one lies outside action lines$) = (0·977)^{r-1} \times 0·023$;
mean $= 1/0·023 = 43·5$

5. $(0·046)^2 = 0·0021$; $(0·160)^2 = 0·026$

6. Remove samples 11 and 12; revised values: $\bar{x}$-chart: $15·2 \pm 2·5$, R-chart: $3·5$. $R_{0·001} = 9·0$; tolerance limits: $15·2 \pm 5·1$.

Author Index

Subject Index

Confidence interval (cont.)
  for slope of line 175
Confidence limits 127
Confounding 278
Consistent estimator 119
Consumer's risk 291
Contingency table 152
Continuity correction 95
Continuous
  data 20
  distribution 64, 81
Contour diagram 281
Contrast 272
Control chart 299
Control limits 300
Controlled variable 167
Correlation coefficient 185–90
  interpretation of 196
  theoretical 195
Covariance 188
Critical path scheduling 347
Critical value 140, 158
Crossed design 277
Cumulative distribution function 83
Cumulative frequency diagram 26–7
Cusum chart 306–12
Curve-fitting 167–71

Data
  continuous 20
  discrete 20
  presentation of 20–34
Decision interval 310
Degrees of freedom 116, 121, 229
  for χ^2 distribution 116, 145, 149, 155
  for F-distribution 155
  for t-distribution 130
  in analysis of variance 238, 248, 266, 274
  of s^2 117, 121, 145
Dependent variable 167
Design of experiments 185, 224, 260
Discrete
  data 20
  distribution 56, 64
Distribution function 83
Distribution
  binomial 57–68
  chi-square, *see* Chi-square distribution

Distribution (cont.)
  continuous 64, 81
  discrete uniform 115
  exponential 98–103, 323
  F- 155–7
  gamma 333
  geometric 80, 102, 316
  hypergeometric 80, 291
  marginal 76
  normal, *see* Normal distribution
  Poisson, *see* Poisson distribution
  t- 129–31, 333
  uniform 83, 87
  Weibull 327
Distribution-free tests 157
Double sampling 293
Duncan's multiple range test 243
Dynamic programming 347

Efficiency 118
Error
  of type I 159, 162
  of type II 159, 162
Estimate or estimator 29, 118
  consistent 119
  interval 107, 126
  point 107, 121
  unbiased 32, 118–19
Events 39
  dependent 42–6
  independent 42–6
  joint 44
  mutually exclusive 40
Evolutionary operation (Evop) 285
Expectation or expected value 66–8, 86, 107–11
Experimental unit 225
Exponential distribution 98–103, 323
Exponential smoothing 313

  relationship with χ^2 distribution 334
F-test 155–7
Factor 225, 257
Factorial experiment, *see* Complete factorial experiment
Factorial n 50
Feedback control system 312
Fixed effects model 276
Fractional factorial design 278

SOCIAL SCIENCE LIBRARY

Manor Road Building
Manor Road
Oxford OX1 3UQ
Tel: (2)71093 (enquiries and renewals)
http://www.ssl.ox.ac.uk

WITHDRAWN

This is a NORMAL LOAN item.

We will email you a reminder before this item is due.

Please see http://www.ssl.ox.ac.uk/lending.html
for details on:

- loan policies; these are also displayed on the notice boards and in our library guide.

- how to check when your books are due back.

- how to renew your books, including information on the maximum number of renewals. Items may be renewed if not reserved by another reader. Items must be renewed before the library closes on the due date.

- level of fines; fines are charged on overdue books.

Please note that this item may be recalled during Term.